W9-BKS-576

EARTH-SHATTERING EVENTS

ANDREW ROBINSON

EARTH-SHATTERING EVENTS

EARTHQUAKES, NATIONS AND CIVILIZATION

Thames & Hudson

For Brian Fagan
Writer, scholar and friend

Frontispiece: San Francisco's City Hall after the earthquake in 1906.

First published in 2016 in hardcover in the United States of America by Thames & Hudson Inc., 500 Fifth Avenue, New York, New York 10110

thamesandhudsonusa.com

Library of Congress Catalog Card Number 2015953647

ISBN 978-0-500-51859-5

Printed and bound in India by Replika Press Pvt. Ltd.

CONTENTS

INTRODUCTION: EARTHQUAKES AND HISTORY 6

1 **EARTHQUAKES BEFORE SEISMOLOGY** 24

2 **THE YEAR OF EARTHQUAKES:** LONDON, 1750 42

3 **THE WRATH OF GOD:** LISBON, 1755 56

4 **BIRTH OF NATIONS:** CARACAS, 1812 74

5 **SEISMOLOGY BEGINS:** NAPLES, 1857 88

6 **ELASTIC REBOUND:** SAN FRANCISCO, 1906 106

7 **HOLOCAUST IN JAPAN:** TOKYO AND YOKOHAMA, 1923 124

8 **BIRTH PANG OF A NEW CHINA:** TANGSHAN, 1976 142

9 **GRIEF AND GROWTH IN THE LAND OF GANDHI:** GUJARAT, 2001 160

10 **WAR AND PEACE BY TSUNAMI:** THE INDIAN OCEAN, 2004 174

11 **MELTDOWN AND AFTER:** FUKUSHIMA, 2011 190

CONCLUSION: EARTHQUAKES, NATIONS AND CIVILIZATION 208

Appendix: Chronology of Earthquakes 235

Map 236

Notes and References 238

Bibliography 244

Acknowledgments 248

Sources of Illustrations 248

Index 249

INTRODUCTION

EARTHQUAKES AND HISTORY

The Colosseum in Rome. About half of its external wall was destroyed by an earthquake, probably in 1349.

California, with its famous San Andreas fault, used to be 'America's earthquake capital'.[1] But recently it has been overtaken by Oklahoma, noted an alarmed editor-in-chief of *Science*, herself a geophysicist and former director of the US Geological Survey (USGS), in 2015.

Before 2000, earthquakes in Oklahoma – even small ones – were uncommon events; indeed the state had long been reputed for its geological stability and absence of major faults. By 2008, though, Oklahoma was experiencing every year an average of one to two earthquakes of magnitude 3.0 or greater, that is, large enough to be felt. (For comparison, the collapse of the Twin Towers in New York in 2001 registered earthquakes of magnitude 2.1 and 2.3.) In 2009, however, there were twenty such earthquakes; in 2010, forty-two of them. The following year, an earthquake of magnitude 5.6 injured two people in the town of Prague, Oklahoma, and destroyed at least sixteen houses plus a turret on a historic university building in nearby Shawnee. During 2014, the number of earthquakes of magnitude 3.0 or greater rose to 585, nearly triple the rate of California and equivalent to more than a century's worth of normal Oklahoman earthquakes. A quake of magnitude 4.2 shook the town of Cushing, a major trading hub for crude oil known as the Pipeline Crossroads of the World, where 54 million barrels were stored underground. Clearly, something unprecedented and potentially dangerous was going on beneath the state. Was nature preparing for an Oklahoman 'Big One'?

The vast majority of scientists were soon convinced that the Oklahoman earthquakes were not natural, like earthquakes in California, but induced – in other words, man-made.

Geologists and seismologists knew that in the early 1960s a series of earthquakes had occurred near Denver in Colorado, where hitherto the natural seismicity had always been low. Between April 1962 and September 1963 seismographic stations near Denver registered more than 700 epicentres with magnitudes of up to 4.3. Then there was a sharp decline in seismicity during 1964, followed by another series of quakes during 1965. It turned out that the US Army was injecting contaminated water from weapons production at its Rocky Mountain arsenal northeast of Denver into a deep well, bored to a depth of about 3,660 metres (12,000 feet).

Injection of the water began in March 1962 and ceased in September 1963 for a year. It resumed in September 1964 and finally ceased in September 1965. Alarmed residents of Denver succeeded in stopping the army's method of disposal and halting the earthquakes.

With the knowledge from this unplanned experiment in mind, in 1969 the USGS designed an experiment at a disused oil field in Rangely, western Colorado. Using existing oil wells, water was injected into a well or pumped out and the pore pressure of the crustal rock (that is, the pressure of the fluid absorbed by the rock) was measured. At the same time an array of seismographs, specially installed in the area, monitored seismicity. There turned out to be an excellent correlation between higher fluid pore pressure and increased seismicity.

The process of injecting fluid into boreholes drilled by the oil and gas industry is now familiar to the public as 'fracking', that is, hydraulic fracturing of shale rock by pumping water, mixed with chemicals and sand, into a shale formation so as to force out trapped natural gas.

Fracking has been demonstrated to cause micro-earthquakes (too small to be felt) and a few felt earthquakes. In Oklahoma, however, fracking is definitely *not* the culprit. Instead, the cause of the earthquakes is the 'dewatering' of oil from wells abandoned as uneconomical in the 1990s but subsequently restarted with the rising price of oil. The problem is that for each barrel of dewatered oil, these wells produce an average of about ten barrels of salt water, that is, 1,600 litres (350 gallons). This large volume of wastewater is being disposed of by injecting it back deep into the ground, where it enters largely unknown geological faults and frequently induces earthquakes. The deeper the injection, the more likely the water will trigger earthquakes. Similar problems have been reported from other states with wastewater-disposal wells: Arkansas, Colorado, Kansas, Ohio and Texas. Outside the US, oil and gas producers in Canada, China and the United Kingdom have also reported such induced seismicity, along with earthquakes induced by geothermal activities in Germany, Switzerland and elsewhere. 'To a large extent, the increasing rate of earthquakes in the mid-continent is due to fluid-injection activities used in modern energy production', declared twelve scientists – including one from the Oklahoma

Geological Survey (OGS) – in a joint paper published in *Science* in 2015.[2] At the same time, a concerned geology professor at Oklahoma State University told the *New Yorker* magazine: 'As scientists, we *knew* the Dust Bowl was going to happen; it wasn't a surprise. It could have been prevented, but scientists failed to effectively communicate what they knew to the people. I don't want that to happen again.'[3] In 2015, the USGS for the first time included induced seismicity in its seismic hazard maps, covering Oklahoma and surrounding states.

Despite this growing scientific consensus, the drilling of disposal wells remained practically unregulated in Oklahoma. No well was denied a permit by the Oklahoma Corporation Commission on grounds of seismicity, nor was injection of the more than 4,600 existing wells curtailed or shut down, unlike in other US states. When the Oklahoma state legislature officially examined the earthquake problem in 2014 and took evidence from local scientists, its report ignored their evidence for induced seismicity, along with copious published scientific evidence from other oil-producing regions, and preferred to cite a local legislator's speculation that the seismicity might be caused by the state's drought. Even the OGS, in its official statements, did not accept that there was sufficient evidence to link the earthquakes to disposal wells, and claimed that the interpretation that best fitted the data for seismicity and fluid injection was 'natural causes'.[4] Only in 2015 did the overwhelming scientific evidence at last compel the state government to introduce some restrictions on the depth and injection rate of disposal wells located within 10 kilometres (6 miles) of the sites of earthquake swarms or quakes of magnitude 4.0 or greater.

Such wilful blindness to science unquestionably has much to do with the power and influence of the oil and gas industry in Oklahoma, which is said to provide one in five of the state's jobs, directly or indirectly, not to speak of its rags-to-riches mythology of 'gushers' discovered by 'wildcatters' going back more than a century. When homes were destroyed by the Prague earthquake in 2011, some of the home-owners refused to speak up, out of deference to the town's well-respected local energy company.

Not only is the oil and gas industry vital to the state's economy, it also funds much of its education, sport and culture, for example the University of Oklahoma, which houses the OGS in the basement of its fifteen-storey earth sciences building. The rest of this building has statues, and a 'well-manicured' garden nearby, dedicated to the achievement of the 'wildcatters' of the oil and gas industry, as noted by a visiting reporter from the science journal *Nature*.[5] At a private meeting in the university in 2013, its president, together with a billionaire oil man, Harold Hamm – the thirteenth child of an Oklahoman sharecropper, whose company had donated more than $30 million to the university – personally pressurized the state seismologist at the OGS not to give public support to a scientific link between the earthquakes and the industry. Hamm's view, as stated in 2014 after a US congressional hearing, is that the earthquakes are 'certainly not related to oil and gas activity'.[6]

Yet, there is more to this story than a clash between science, business and government, of a kind familiar from the current rancorous US debate over climate change. Oklahoma's geology has created both wealth, in the form of oil and gas, and hazard, in the form of induced earthquakes. Many Oklahomans, whatever their level of income, appear to consider this opportunity for economic prosperity worth the seismic risk.

Their contemporary Faustian bargain with earthquakes – in this case man-made quakes – is one small episode in the long and fascinating history of man's relationship with seismicity. 'People don't like earthquakes, and yet, over and over again, people choose to live in areas susceptible to earthquakes', note two well-known US seismologists, Susan Hough and Roger Bilham, in their historical survey, *After the Earth Quakes*.[7] For the ancient Greeks, Romans, Hebrews and Persians, the Chinese and the Japanese, the Maya and the Incas, and many other peoples, earthquakes were an accepted part of life. From antiquity until the present day, on every continent, civilizations have deliberately accepted the risk of periodic seismic destruction.

* * *

In the mid-19th century, as the study of earthquakes was slowly becoming a science, Charles Darwin, the great English naturalist who originally made his name as a geologist, experienced a severe earthquake on the coast of Chile while he was circumnavigating the planet in His Majesty's ship *Beagle*. In his classic travel diary, generally known as *The Voyage of the Beagle*, Darwin ranked the earthquake and its impact as the most 'deeply interesting' sight of his entire five-year journey.

For Darwin, born and brought up in geologically stable England, this was his first personal encounter with the earth's instability. As he beheld the newly devastated Chilean city of Concepción in 1835, Darwin brooded pessimistically that:

> Earthquakes alone are sufficient to destroy the prosperity of any country. If, for instance, beneath England, the now inert subterranean forces should exert those powers which most assuredly in former geological ages they have exerted, how completely would the entire condition of the country be changed! What would become of the lofty houses, thickly packed cities, great manufacturies, the beautiful public and private edifices? If the new period of disturbance were first to commence by some great earthquake in the dead of night, how terrific would be the carnage! England would at once be bankrupt; all papers, records, and accounts would from that moment be lost. Government being unable to collect the taxes, and failing to maintain its authority, the hand of violence and rapine would go uncontrolled. In every large town famine would be proclaimed, pestilence and death following in its train.[8]

Thankfully Britain has never been put through Darwin's imagined seismic ordeal. Not that even Britain – including its capital London – has been earthquake-free: as recently as 2008, a magnitude-5.2 earthquake caused damage to chimney-stacks, roofs and garden walls, and one serious injury, as reported in the national press. Over the centuries, there have been dozens of British earthquakes, catalogued in Charles Davison's *A History of British Earthquakes*.

One in 1248 threw down the vaulted ceiling of Wells Cathedral. Another in 1580 caused part of the white cliffs at Dover to fall into the English

Channel, killed two children in London, rang the great bell in the Palace of Westminster, and is thought to have influenced William Shakespeare's play *Romeo and Juliet*, when Juliet's nurse remembers an unforgettable day:

> 'Tis since the earthquake now eleven years
> And she was wean'd – I shall never forget it –
> Of all the days of the year upon that day.[9]

The worst one, in 1884, wrecked houses and toppled churches in and around the ancient Roman town of Colchester, while pitching the engine driver of a waiting express train to London out of his cab onto the station platform. It rattled London, too. In the Houses of Parliament, within the Palace of Westminster, puzzled MPs were 'stopped in their tracks, jolted against walls, or felt papers and briefcases jerked from their hands'.[10] Officials were immediately despatched to the cellars of the palace to investigate the possibility that there had been a Guy Fawkes-style explosion, perhaps set off by the notorious Dynamiters who were at that time being prosecuted by the police for their Irish nationalist activities. Fortunately, the shaking lasted for a mere five seconds, with an estimated magnitude of 4.6.

But the most historically significant British earthquakes were undoubtedly those of 1750, the so-called 'Year of Earthquakes', which struck both London and elsewhere in the country. Although they induced the usual panic in the public and righteousness among religious preachers, they also marked a new beginning: the objective study of earthquakes, as reported and analysed at length by the fellows of the Royal Society, then probably Europe's leading scientific organization. Strange to say, earthquake science started in Britain in 1750. These earthquakes therefore, despite their limited damage, deserve a chapter in this book of their own (see Chapter 2).

Of course, no British earthquake has come close to the magnitude of the earthquakes that strike continental European countries, notably Greece, Italy, Portugal and Rumania, and, further afield, Algeria, the Caribbean islands, Chile, China, Colombia, India, Indonesia, Iran, Israel, Japan, Mexico, Morocco, Nepal, New Zealand, Pakistan, Peru, Russia, Taiwan,

Turkey and of course the United States. Not only on its Pacific west coast (around San Francisco and Los Angeles), and in Alaska, but also on its Atlantic east coast (around Boston and Charleston) and even in mid-continent: Missouri experienced an earthquake so powerful that it briefly reversed the course of the Mississippi River in 1812.

In Chile, Concepción has been struck some ten times since its founding in 1550, most recently in 2010 by a magnitude-8.8 earthquake, the sixth largest ever to be recorded by a seismograph. Since the earthquake magnitude scale is not linear, but rather logarithmic, the energy of a high-magnitude earthquake is far greater than one might expect from its magnitude number. Thus, the energy released by a magnitude-8.0 earthquake is about 32 times more than one of magnitude 7.0, and 32 squared (approximately 1,000) times more than one of magnitude 6.0. The most powerful earthquake ever recorded – of magnitude 9.5 in Chile in 1960 – released more than 20,000 times the energy of the atomic bomb dropped on Hiroshima in 1945, and about a quarter of the entire seismic energy release of the planet since the beginning of the 20th century. Rupturing 1,000 kilometres (650 miles) of fault running down Chile's coastline, the 1960 earthquake was so powerful that 'it wobbled the planet' – and also set seismologists thinking about how to devise a universally applicable magnitude scale (used throughout this book), to replace the 'Richter' magnitude scale devised in the 1930s by seismologist Charles Richter for measuring moderate earthquakes in southern California with a specific, outdated seismograph.[11] Western South America, measured solely by the magnitude of its earthquakes and their energy release, rates as 'the most seismically active region in the world'.[12]

Was Darwin correct about the economic fragility of earthquake-prone regions? Clearly not, when we consider the long periods of prosperity in many of the above-listed countries. At the present time, the United States has the world's largest economy, China the second, Japan the third. China and Japan are among the world's most seismically active countries, however seismicity is measured: whether according to contemporary seismographic monitoring, the historical earthquake record or the number of earthquake fatalities. Some 22 per cent of the world's earthquakes of magnitude 6.0 or greater occur in Japan.

On this evidence, we might even argue that destructive earthquakes, for all their horrors, can enhance economic growth over the longer term. 'Earthquakes create a lot of business', remarked the so-called 'father of seismology', John Milne, just over a century ago.[13]

Certainly economic growth resulted from the destruction of San Francisco by an earthquake and fire in 1906 (see Chapter 6). Following a period of reconstruction, San Francisco went on to flourish and in the 1950s give birth to the high-tech industrial area on the San Andreas fault southeast of San Francisco now known as Silicon Valley. 'It is conventional, and by no means inappropriate, to think of disasters in strictly negative terms, but calamities have also often presented opportunities', writes historian Kevin Rozario in *The Culture of Calamity: Disaster and the Making of Modern America*. 'Americans, especially those in positions of power and influence, have often viewed disasters as sources of moral, political and economic renewal.'[14]

Indeed, natural disasters can be powerful promoters of corporate and free-market interests, argues social activist Naomi Klein in *The Shock Doctrine: The Rise of Disaster Capitalism*, with specific reference to the Indian Ocean earthquake-induced tsunami in 2004. Some thinkers have gone so far as to see earthquakes as blessings in disguise. After the destruction of Lisbon in 1755, the philosopher Immanuel Kant claimed: 'Just as we complain of ill-timed or excessive rain, forgetting that rain feeds the springs necessary in our economy, so we denounce earthquakes, refusing to consider whether they too may not bring us good things.'[15] In 1848, in his *Principles of Political Economy*, John Stuart Mill predicted long-term benefits from such disasters, because they obliterated old stock and encouraged manufacturers to introduce efficiency savings in production processes. From a religious and political perspective, Mahatma Gandhi maintained that a great earthquake in north India and Nepal in 1934 was a warning to caste Hindus against the sin of Untouchability. 'Visitations like droughts, floods, earthquakes and the like, though they seem to have only physical origins, are, for me, somehow connected with man's morals', Gandhi publicly announced.[16]

Since prehistory, human societies have cohabited with seismicity in a 'fatal attraction' (the evocative phrase of geophysicist James Jackson), because the advantages of living with earthquakes easily outweigh the disadvantages.[17] More than half of the world's largest cities – as many as sixty of them – lie on plate-tectonic boundaries such as the San Andreas fault, in areas of major seismic activity. They include Ankara, Athens, Beijing, Cairo, Caracas, Delhi, Hong Kong, Istanbul, Jakarta, Karachi, Lisbon, Lima, Los Angeles, Manila, Mexico City, Naples, Osaka, Rome, San Francisco, Santiago, Shanghai, Singapore, Taipei, Teheran and Tokyo. Some of them – notably Caracas, Lisbon, Lima, Los Angeles, Manila, Mexico City, Naples, San Francisco, Teheran and Tokyo – have suffered major destruction from earthquakes during the past two or three centuries.

Plate-tectonic boundaries often coincide with coastlines and islands, which have always provided fruitful environments for human settlement – as in California, Chile, Greece, Indonesia, Italy and Japan. At present, the vast majority of Chile's population live in a narrow but fertile strip of land between the Pacific coast on the west and the Andes Mountains in the east, which is a dangerous subduction zone, geologically speaking, in which the Nazca plate of the eastern Pacific pushes its way eastwards and subducts – dives down – beneath the stationary South American plate, thereby generating great earthquakes and thrusting up the Andes. In antiquity, the Greeks and Romans – notwithstanding frequent earthquakes in the Aegean area and in the Italian peninsula – created colonies and empires and enduring monuments: one of which, the Colosseum in Rome, built in the late 1st century AD, stands half-ruined by an Italian earthquake (probably a major one in 1349). In prehistory, some 2 million years ago, tectonic movements in the Dead Sea fault system of Palestine produced a lush and inviting valley in the midst of arid desert that attracted mankind's earliest emigrants from Africa, who eventually created cities such as Jericho, one of the oldest in the world, dating back to the 7th millennium BC.

Important cities destroyed by earthquakes and their subsequent fires have proved to be extraordinarily resilient – unlike villages, which tend to be abandoned or relocated. Indeed, Jericho supports this observation. According to a famous passage in the Bible, Joshua and his people are said to have passed over the flooded River Jordan by unknown means, laid siege to Jericho and captured it on the seventh day after blowing their ram's-horn trumpets and giving a great shout, which made the city's walls suddenly fall down flat. Almost certainly, Joshua was the beneficiary of an earthquake followed by a landslide, which between them dammed the Jordan and flattened Jericho. Archaeological excavation of the site of Jericho shows that it has been devastated by multiple earthquakes over many centuries; such landslides in the Jordan valley were reported in historical times in AD 1160, 1267, 1534, 1546, 1834 and 1927. An earthquake at Jericho in 31 BC damaged the palace of Herod the Great and provoked the king to reassure his troops that the cause was natural, not divine; the latest one, in 1927, almost totally demolished the modern city. Yet, after each seismic shaking, Jericho was rebuilt.

In classical antiquity, Pompeii, near earthquake-prone Naples, was devastated by an earthquake in AD 62 or 63. The Roman emperor Nero, after visiting the city, recommended that the seismic damage was so bad that Pompeii should be abandoned. But it was instead rebuilt – just in time for Pompeii's permanent destruction by the volcanic eruption of Vesuvius in 79. The trading and pleasure city of Antioch (modern Antakya) on the Mediterranean coast of southeastern Turkey was ravaged by earthquakes in AD 115, 458, 526 and 528, the first of which injured the Roman emperor Trajan, who was compelled to shelter in the city's circus. Antioch, considered in its time to be comparable with Athens, Rome, Alexandria and Constantinople – with a population of perhaps half a million people in the first two centuries AD – was always rebuilt: as many as fifteen times over the past twenty-three centuries. In Persia/Iran, the site of Teheran was damaged or completely destroyed by earthquakes in the 4th century BC, AD 855, 958, 1177 and 1830.

In the modern period, Lisbon was rebuilt after the cataclysm of 1755, as were Tokyo and Yokohama after the Great Kanto earthquake in 1923. In China, the industrial centre of Tangshan was rebuilt after a night-time earthquake in 1976 killed as many as 750,000 sleeping Chinese (while sparing all but seventeen of Tangshan's 10,000 miners working underground). In Central America, the old capital of Guatemala, Antigua, was ruined and rebuilt four times from 1586 in less than 300 years; the capital of Nicaragua, Managua, ten times in less than 200 years. In fact, in recorded history no city has ever been abandoned as a result of a great earthquake, except for Port Royal in Jamaica, two-thirds of which slid under the Caribbean Sea after an earthquake in 1692, thereby drowning and suffocating some 25,000 of Port Royal's inhabitants in water and sand.

So, earthquake-prone cities – including many capitals – generally recover from seismic catastrophes and frequently prosper. What about societies and nations? Here, the historical record is less consistent and inevitably open to interpretation, debate and dispute.

At one pole, there is the much-quoted observation attributed to historian Will Durant that: 'Civilization exists by geological consent, subject to change without notice'.[18] At the other, there is the influential opinion of the scientist and geographer Jared Diamond, author of *Guns, Germs, and Steel: The Fates of Human Societies* and *Collapse: How Societies Choose to Fail or Succeed*. In the latter book, Diamond almost ignores natural disasters, and totally ignores earthquakes and volcanic eruptions. Both Durant and Diamond are too extreme, in my view. While societies and nations are unquestionably more likely to fail or succeed as a result of human activities, such as warfare or empire-building, they may also be destabilized or developed by great natural forces, such as floods or earthquakes; moreover sciences, such as seismology, can to some extent tip the balance in favour of success against the forces of nature. The problem – and the subject of this book – is, of course, to understand exactly how human agency and great earthquakes have interacted, not only in the short term, but also in the long perspective of history.

Consider a much-debated example from antiquity. Around 1200 BC, there was a catastrophic, apparently simultaneous, collapse of the Bronze Age cultures at an astonishing forty-seven archaeological sites around the eastern Mediterranean, including Mycenae (mainland Greece), Knossos (Crete), Troy (Anatolia) and Armageddon (the Levant). The Bronze Age civilization in Greece was succeeded by a Dark Age, which was apparently illiterate. This lasted for some four centuries until the appearance of Homer's poetry in the 8th century BC, along with the Greek alphabet.

Could earthquakes have been responsible for this collapse? Possibly. Seismologists are certain that the outer walls of Mycenae, directly beneath its celebrated Lion Gate, were built on top of a fault scarp, which must have been created during a major earthquake. At Knossos, the chief excavator from 1900 until his death in 1941, Arthur Evans, experienced a local Cretan earthquake while digging. At Troy, the excavations were shaken by a major earthquake in 1912. Both Evans and a key excavator of Troy in the 1930s, Carl Blegen, were sympathetic to a seismic interpretation of the Minoan and Trojan archaeological evidence.

However, many current archaeologists are unconvinced. Not being geologically trained, archaeologists tend to miss evidence for seismic damage; and even when the seismic evidence is too plain to ignore, they are inclined to dismiss earthquakes as events with profound ramifications. Most of them prefer to attribute the decline or collapse of ancient societies to war, invasion, social oppression, economic corruption, environmental abuse and so on – rather than natural disasters, 'acts of God'. The idea that a natural disaster might on occasion operate in tandem with human agency – as with the possible earthquake and landslide near Jericho and Joshua's subsequent capture of Jericho – is seen by these archaeologists 'as a capitulation, a sign of a weak theory that must be bolstered by unlikely coincidences', writes a geophysicist and palaeoseismologist, Amos Nur, in *Apocalypse: Earthquakes, Archaeology, and the Wrath of God*.[19]

In Nur's view, by contrast, 'When many similarly oriented walls at a site have fallen in the same direction', as at Jericho, Mycenae and Troy, 'particularly when they have buried grain, gold or other valuables in their fall, the action of an army is an unlikely cause.'[20] While Nur fully recognizes the ambiguous nature of the geological, archaeological and literary evidence for ancient earthquakes – including the many references to them in classical Greek drama (notably the plays of Euripides) and in the Bible (where an earthquake accompanies both the crucifixion and the resurrection of Jesus Christ) – he nonetheless makes a strong case for their importance.

In truth, neither earthquakes on their own nor invasions on their own will account for the Bronze Age collapse around 1200 BC. A more plausible, if less than wholly satisfying, natural-cum-human explanation could be as follows, as argued by classicist Eric Cline in *1177 B.C.: The Year*

Civilization Collapsed. Initially, all of these ancient societies (Mycenae, Knossos, Troy, Armageddon and so on) were weakened not by one giant earthquake and its aftershocks but by a long sequence of earthquakes, in which one earthquake triggered another during the period 1225–1175 BC. This possibility – dubbed an 'earthquake storm' by Nur and others – is supported (on a geological, rather than human, time-scale) by a series of historically attested major earthquakes that struck the eastern Mediterranean area in the middle of the 4th century AD: for example, Sicily, Constantinople and Jerusalem/Petra were each struck in different months of AD 363. Again, during the 20th century AD, there was an exceptionally high level of seismicity in the eastern Mediterranean area, as measured by seismographs between about 1900 and 1980: Turkey alone experienced thirty-two earthquakes of magnitude greater than 6.0. Subsequently, during and after the earthquake storm, suggests Cline, the weakened Bronze Age societies were destroyed by human agency, including seaborne invasions by various groups of marauders, whom archaeologists generally designate as the Sea Peoples.

Earthquakes of the pre-modern world, where historical evidence is thin, occupy Chapter 1 of the book. Then we move on to the modern period (in Chapters 2–11) – including the development of earthquake science after 1750 (Chapters 2, 5 and 6) – where records are comparatively plentiful. Here it is clear that although earthquakes have not had the power to break or make states and civilizations, from time to time they have altered the course of history and determined the fate of nations.

Consider the following great earthquakes of the past two and a half centuries.

In Portugal, the devastation of its capital, Lisbon, by an earthquake in 1755 (see Chapter 3) accelerated the long-term decline of the country in both Europe and the colonial world, caused by its over-reliance on gold revenues from its colony Brazil and the pernicious influence of Jesuit religious orthodoxy. Although the Jesuits were expelled and Lisbon was gradually, and impressively, rebuilt under the near-dictatorship of the marquis of Pombal, the country was economically weakened, especially after Brazil gained its independence in 1822. In Europe as a whole, political, religious, philosophical and scientific thought were significantly changed by Voltaire's lacerating writings about the Lisbon earthquake. 'By striking at a time when there was a particularly delicate balance of power between church and state, and between science and religion,' notes Nur, 'the earthquake tipped the scales and changed society around the world.'[21]

In Latin America, an earthquake in Venezuela in 1812 (see Chapter 4) destroyed much of the country's buildings including those of its capital, Caracas. The damage happened to be worst in the areas controlled by Simón Bolívar's recently proclaimed First Republic of Venezuela and relatively light in areas sympathetic to the colonial ruler, Spain: a fact immediately exploited by the local Catholic authorities, who supported Spain. By Bolívar's own admission, the earthquake directly precipitated the republic's collapse four months later under attack by Spanish forces, which captured Bolívar and sent him into exile. There he unexpectedly became the leader of a much wider independence movement than the one he had led in Venezuela before the earthquake. Indirectly, therefore, the 1812 earthquake may be said to have led to Bolívar's liberation of Bolivia, Colombia, Ecuador, Peru and Venezuela from Spanish rule in the 1820s.

In Japan, the Great Kanto earthquake in 1923 (see Chapter 7), which lasted for a crippling five minutes, struck as the midday meal was being cooked. The firestorms it ignited left two-thirds of Tokyo and four-fifths of Yokohama in ashes, and cost at least 140,000 lives, including those of many Korean immigrants murdered by Japanese vigilantes. Martial law was required to control the chaos, giving a new degree of authority to the army. The massive cost of rebuilding the cities between 1923 and 1930 created an economic stress and a financial panic. These events together set the scene for a more authoritarian imperial government in 1927, which favoured military intervention in China. Japan's invasion of China (Manchuria) in 1931, along with the worldwide economic depression of the 1930s, led to a pervasive militarization of Japanese society, and eventually to Japan's entry into the Second World War in 1941.

In China, the appalling Tangshan earthquake in 1976 (see Chapter 8) literally shook the deathbed of Chairman Mao Zedong in not-so-far-off Beijing, and figuratively shook up the Communist Party leadership. Though only 23 seconds in duration, its death toll was the highest for any 20th-century earthquake. Mao's death, just over a month after the disaster, prepared the way for the leadership of Deng Xiaoping from 1978 and the subsequent transformation of China into a world economic power. While the death of Mao was the proximate cause of these pivotal changes, the Tangshan earthquake can be regarded as their catalyst. For the Chinese government's incompetence in dealing with the Tangshan catastrophe exposed Mao's Cultural Revolution as a sham, and undermined the Chinese people's faith in its Maoist government to protect them.

In India, the destruction of towns and cities in the state of Gujarat by an earthquake near Bhuj in 2001 (see Chapter 9) led to the forced resignation of the state's chief minister nine months later, after he had failed to begin effective reconstruction. The next chief minister, Narendra Modi – a Hindu nationalist appointed without an election – responded to the destruction by launching a rapid, uncontrolled, industrialization of Kutch, the area of western Gujarat affected by the earthquake. While consolidating Modi's power base, this economic regeneration also appeared to offer a model for the development of other Indian states. In 2014, during India's national elections, the controversial Modi was easily elected as India's prime minister, largely on the expectations aroused by his economic record in Gujarat, predicated on the destruction caused by the earthquake.

In the Indian Ocean, a massive submarine earthquake off the coast of Sumatra generated a tsunami in 2004 (see Chapter 10), which caused mayhem on the coasts around the Indian Ocean and the loss of about 230,000 lives. The worst affected countries were Indonesia, in particular the province of Aceh in Sumatra, and Sri Lanka, especially its northern and eastern coasts, which are a Tamil-majority region. In both of these areas, a local armed insurgency had long been fighting the country's central government. But whereas in Aceh the tsunami disaster led directly to an enduring peace treaty between the Free Aceh Movement and the Indonesian government, in Sri Lanka the effect was the opposite: the disaster solidified the grip of the Sinhalese nationalist government in Colombo, which went on to annihilate the separatist movement led by the Tamil Tigers with a concerted military attack on the northeast of the island in 2009.

As for the Great East Japan (Tohoku) earthquake in 2011 (see Chapter 11), and the tsunami that overwhelmed the Fukushima Daiichi nuclear power plant, this disaster was described by the then Japanese prime minister as 'the biggest crisis' Japan had encountered since the end of the Second World War.[22] It is too soon to say what its long-term effects will be. However, the earthquake's jolt to the Japanese political system is already evident in the emergence of a nationalist central government at the same time as the rise of stronger local government, especially in the tsunami-affected northeast, and a vigorous national volunteering movement. There have also been worldwide reverberations in the nuclear power industry, as the Japanese government grapples with the clean-up of the wrecked power plant, which is expected to require decades.

Overall, therefore, history suggests that great earthquakes have indeed sometimes been important in the decline, collapse and rebirth of societies. Darwin was right to draw attention to their awesome power in 1835. But he was probably wrong to suggest that mid-19th-century England, given its strong government and industrial and financial resources, not to speak of its extensive colonial empire, would have struggled to rebound from such a hypothetical geological assault. For similar reasons, 20th-century San Francisco and Tokyo rebounded relatively fast from devastating earthquakes. Compare the effects of two major earthquakes in 2010, which were less powerful than those in San Francisco in 1906 and Tokyo in 1923, but still highly destructive. One of them, of magnitude 7.1, struck New Zealand, 40 kilometres (25 miles) from the city of Christchurch, yet caused not a single fatality. The other, of magnitude 7.0, struck Haiti, 25 kilometres (16 miles) from its capital Port-au-Prince, and caused somewhere between 85,000 and 316,000 deaths; the higher figure is the Haitian government's estimate, which is disputed by international aid agencies. The most significant reason for the huge difference in the fatalities in Christchurch and Port-au-Prince was the reinforced construction of buildings in New Zealand, as compared with the unreinforced construction of buildings in Haiti – a fact that of course depends on the very different degree of political, economic, technological and scientific development of New Zealand and Haiti. Where government is weak and resources are poor, Darwin's pessimism may be justified.

The long-term impact of a great earthquake depends on its epicentre, magnitude and timing – and also on human factors: the political, economic, social, intellectual, religious and cultural resources specific to a region's history. As we shall now discover, each earthquake-struck society offers its own particular lesson; and yet, taken together, such earth-shattering events have important shared consequences for the world.

CHAPTER

I EARTHQUAKES BEFORE SEISMOLOGY

The Kashima Deity Napping, detail, Japanese woodblock print, *c.* 1855. When the deity is off guard, a mischievous underground catfish (*namazu*) provokes earthquakes/fires.

The further back in time one looks, the harder it is to detect an earthquake and its signature. Disappointingly, there is almost no clear account of seismicity in some 3,500 years of ancient Egyptian inscriptions and papyri, beginning around 3000 BC, because Egypt has low seismic activity. The first reliable reports of earthquakes date only from 780 BC in China, 464 BC in Greece, 461 BC in Italy, AD 599 in Japan and as late as 1567 in the Americas (in Mexico), although there is a report in the ancient Chinese *Bamboo Annals* of the shaking of Taishan mountain in Shandong province in 1831 BC.

Archaeology provides further, less definite, evidence for earthquakes in earlier periods. Dholavira, a site in Gujarat belonging to the Indus civilization dating from the late 3rd millennium BC, shows slip faults in sections and displacement of architectural features, subsequently repaired by the town's occupants, suggesting an ancient earthquake. A Minoan temple at Anemospilia on Crete, not far from Knossos, destroyed by fire around 1700 BC, yielded the skeletons of what appeared to be a priest and his female acolyte and, on the temple's altar, a young man lying on his side, probably once bound hand and foot, with a large ceremonial bronze knife lying on his bones. Forensic examination of his burnt skeleton showed that the man's blood had drained from the whitened upper part of his body, but not from the blackened lower part. Apparently, he was in the midst of being killed by the priest when an earthquake struck – possibly as a sacrificial offering in response to foreshocks, intended to avert a greater shock, as the excavators grimly speculated.

Also significant, though even less conclusive, are mythological traditions, which may transmit the experience of earthquakes from still earlier times. The majority of earthquake creation myths involve animals. In Mexico, among the Aztecs, a jaguar leaping towards the sun was thought to cause earthquakes; among today's Tzotzil Maya, they occur when a jaguar scratches itself against the pillars of the world. In British Columbia, the Haida Indians imagined a thunderbird fighting with a whale and dropping the whale from its talons into the Pacific Ocean, causing earthquakes and tsunamis. In India, among the Hindus, one of eight great elephants supporting the earth became weary from time to time, lowered its head and gave an earth-shattering shake. In Mongolia, lamas conceived of a gigantic frog carrying the earth on its back and giving periodic twitches. In Japan, where earthquake mythology was at its most sophisticated, perhaps unsurprisingly, earthquakes came to be symbolized by a dragon, which then morphed into a large cosmic fish, and finally became a mischievous giant catfish, known as a *namazu*, living in the mud beneath the earth. The *namazu* was normally restrained by a deity who protected Japan from earthquakes by keeping a mighty rock on the creature's head. The supposed rock can be seen at Kashima, a place about 100 kilometres (62 miles) from Tokyo that has remained comparatively free from earthquakes. However, the Kashima god occasionally dozed off or had to leave his post in order to confer with other gods. At such times the *namazu* was licensed to twitch its barbels, writhe around and generally play pranks – with disastrous results for human beings on the surface. The mythology is brilliantly and humorously depicted in coloured woodblock prints, known as *namazu-e*, first created after a major earthquake near Edo (modern Tokyo) in 1855. In one print, the restless catfish is seen being attacked by every inhabitant of Edo's entertainment district except for the carpenters and other artisans, who inevitably do well out of earthquake disasters. Nowadays, images of catfish appear in emergency earthquake preparedness programmes in Japan, such as the Earthquake Early Warning logo of the Japan Meteorological Agency.

That said, in some cultures earthquakes take human form. The inhabitants of an island in Indonesia attributed them to a demon, which shook with rage when not propitiated by certain sacrifices. The Maori of New Zealand had an earthquake god, Ruaumoko, who had been accidentally pressed into the ground as a baby when his mother rolled on top of him. Among the historic Quiché Maya, whose modern descendants live in Guatemala, their ancient book of creation, the *Popol Vuh*, features an evil character, Cabracan, whose name means Earthquake in both classical and modern Quiché. His legs provide the pillar (or pillars) that hold up the earth; earthquakes are caused by his movements. In classical Greece, Poseidon, god of the sea, was usually considered to be responsible for earthquakes. Poseidon produced shaking while striking his trident on the ground when he became annoyed. When, in the Trojan War, he joins the Greek side in a battle against Ilion (Troy), Homer's *Iliad* reports that: 'Poseidon made the solid earth quake beneath, and the tall summits of the hills; Mount Ida shook from head to foot, and the citadel of Ilion trembled'.[1]

However, Greek thinkers were the first to propose natural, rather than mythological, explanations for earthquakes. Thales, for example, writing around 580 BC, believed that the earth was floating on the oceans and that water movements were responsible for earthquakes. By contrast, Anaximenes, who also lived in the 6th century BC, proposed that rocks falling in the interior of the earth must strike other rocks and produce reverberations. Thucydides, writing about an Aegean tsunami that occurred in 426 BC, thought that such shocks could drive back the sea, which then violently recoiled and inundated the land. Anaxagoras, during the same century, regarded fire as the cause of at least some earthquakes. Aristotle, in the 4th century, believed in a 'central fire' inside caverns in the earth from which flames, smoke and heat rapidly rose and burst violently through the surface rocks, causing volcanic eruptions. As the subterranean fires burnt away the rocks, the underground caverns collapsed, generating earthquakes. Aristotle even classified earthquakes into types according to whether they shook structures and people in mainly a vertical or a diagonal direction, and whether or not they were associated with escaping vapours.

By the time that the Bible came to be written in its final form, natural and supernatural interpretations of earthquakes were inextricably entwined. Are biblical earthquakes metaphorical or might they refer, in some cases, to geological earthquakes in the Holy Land? As with all attempts at linking biblical events with history, the answer is always controversial.

In the New Testament, at the crucifixion, 'Jesus again gave a loud cry, and breathed his last. At that moment the curtain of the temple was torn in two from top to bottom. There was an earthquake, the rocks split and the graves opened'. Matthew's description may have been inspired by the 31 BC earthquake that damaged the temple in Jerusalem as well as Herod's palace in Jericho. At the resurrection of Jesus from his tomb, 'Suddenly there was a violent earthquake; an angel of the Lord descended from heaven; he came to the stone and rolled it away, and sat himself down on it.'[2] Here the supernatural element evidently predominates, although the shaking of a real earthquake could have dislodged a tombstone by natural means.

In the Old Testament, numerous references to earthquakes are mostly associated with the wrath of God. But in a passage from the book of Kings, in which the prophet Elijah stands on top of Mount Horev, wishing to die after his rejection by the people of Israel, the biblical writer's interpretation is more ambiguous:

> And behold, the Lord passed by, and a great and strong wind rent the mountains, and brake in pieces the rocks before the Lord; but the Lord was not in the wind: and after the wind an earthquake; but the Lord was not in the earthquake: and after the earthquake a fire; but the Lord was not in the fire: and after the fire a still small voice.[3]

Presumably, this was intended to mean that wind, earthquake and fire, though frightening, were natural occurrences, whereas 'God was the whisper in the stillness', comments geophysicist Amos Nur.[4]

The infamous destruction of the cities of Sodom and Gomorrah in the book of Genesis is regarded as punishment for moral transgressions: 'the men of Sodom were wicked, great sinners against the Lord.'[5] Even so, it may have referred to an earthquake, at least in part. No earthquake is explicitly mentioned, but an earthquake and fire seem to be implied by the statement that the Lord 'rained down fire and brimstone from the skies on Sodom and Gomorrah. He overthrew those cities and destroyed all the Plain, with everyone living there and everything growing in the ground.' When, early in the morning, Abraham looked out upon the plain, towards the two cities, he saw 'thick smoke rising high from the earth', as if the land had been incinerated.[6]

Of course, this does not constitute proof that Sodom and Gomorrah ever existed on the ground. However, a statement about the Dead Sea area and Sodom by Strabo, the noted ancient Greek geographer, published in the early 1st century AD, is highly suggestive that they were real places:

> Near Moasada are to be seen rugged rocks that have been scorched, as also, in many places, fissures and ashy soil, and drops of pitch dripping from smooth cliffs, and boiling rivers that emit foul odours to a great distance, and ruined settlements here and there; and therefore people believe the oft-repeated

> assertions of the local inhabitants, that there were once thirteen inhabited cities in that region of which Sodom was the metropolis . . . and that by reason of earthquakes and eruptions of fire and of hot waters containing asphalt and sulphur, the lake burst its bounds, and rocks were enveloped with fire; and, as for the cities, some were swallowed up and others were abandoned by such as were able to escape.[7]

Moreover, Strabo identified certain ruins at the southwestern tip of the Dead Sea as the site of biblical Sodom.

To date, no suitable ruins of Sodom have been located by archaeologists, who have suggested various locations for the biblical city around the Dead Sea. There is, however, considerable consensus among geologists that an earthquake is the most likely explanation for Sodom's destruction. It may have occurred around 2100 BC with a magnitude of at least 6.8, according to geophysicist Ari Ben-Menahem. Also possible is that fault movements may have released petroleum and sulphurous gases from fissures, which then ignited – either spontaneously by lightning or by seismic activity – thereby explaining the biblical reference to brimstone and fire. Such ignition was observed in a petroleum seep from the flank of the San Gabriel Mountains, north of Los Angeles, during the Fort Tejon earthquake in 1857.

A second famous episode of urban destruction from the Old Testament, Joshua's capture of Jericho with his army's trumpet blast, traditionally dated between 1400 and 1250 BC by biblical historians, also lacks a specific reference to an earthquake. But if we put aside miracles, seismic intervention appears even more probable than at Sodom and Gomorrah, as mentioned in the Introduction. Archaeological excavation of the enormous mound known as Tel Jericho indicates as many as twenty-two levels of destruction and the repair or complete rebuilding of the city's walls no fewer than sixteen times. Some of this destruction was almost certainly due to earthquakes, given that devastating earthquakes have undoubtedly occurred at or near Jericho in historical times. In one level, under fallen walls excavators found storage jars full of grain (carbonized by fire), dating from 1600–1550 BC. Normally, valuable grain would have been taken away by departing inhabitants or seized and consumed by invaders – had an earthquake not destroyed the city's walls and buried the grain jars. In another level, dating from around 1400 BC, two adjacent skeletons of people crushed by falling walls are particularly revealing because one skeleton has been decapitated by the later motion of a geological fault that clearly runs – in the form of a crack in the ground – between the torso and the head. Were they earthquake victims?

No single excavation level of Tel Jericho has been identified with Joshua's attack, and this is most unlikely to be feasible, not least because of the extensive damage to the site by earthquakes (and archaeologists). Nor is there a totally convincing explanation for the absence of an 'earthquake' in the biblical account. The most likely hypothesis is that a single earthquake dammed the River Jordan and at the same time demolished the walls of Jericho, as discussed earlier. Thus, 'Joshua would have arrived to find that the city had already been conquered for him, apparently by God', suggests Nur. 'The embellishments of marching around the city seven times, the shout and the sound of trumpets would all make the story more dramatic in the retelling as it was passed down through the years.'[8]

A century or two later, around 1200 BC, came the collapse of the Bronze Age civilizations in the eastern Mediterranean discussed in the Introduction. Here one finds almost no literary sources like the Bible apart from a handful of cryptic inscriptions, only the archaeological evidence of destruction at forty-seven sites ranging from Knossos in Crete and Mycenae and Pylos in mainland Greece through Troy, Miletus and Hattusas in Anatolia, Carchemish, Aleppo and Ugarit in Syria, Megiddo (Armageddon), Lachish and Ashkelon in the Levant, to four less famous sites on the island of Cyprus, such as Enkomi. At some sites, for instance Knossos and Troy, the original excavators attributed the destruction to earthquakes; at others, such as Hattusas and Pylos, they did not. Subsequent studies of a site do not always agree. Even Nur admits that the damage at many sites could have been caused by human hands as easily as by earthquakes. But, he argues, given the indisputable history of seismicity of the region – such as the magnitude-7.4 earthquake at Troy in 1912 that destroyed the former house of the original excavator, Frank Calvert – it is 'indefensible to dismiss earthquakes without serious consideration'.[9]

Sceptical archaeologists ask: if earthquakes destroyed the cities, why were they not rebuilt? Reconstruction was the rule after other ancient earthquakes, observes classicist Robert Drews in *The End of the Bronze Age*. 'One is therefore reluctant to believe that circa 1200 BC a number of the most important places in the eastern Mediterranean were hit by a quake from which they could not recover.'[10] For such earthquake sceptics, the preferred explanation of the Bronze Age collapse is maritime invasions by the Sea Peoples, whose existence is hinted at, but never properly defined, in ancient Egyptian chronicles. The Sea Peoples are mentioned (and depicted as captives) under the mysterious ethnic names, Peleset, Tjekker, Shekelesh, Shardana, Danuna, and Weshesh, in the Egyptian hieroglyphic inscriptions of Ramesses III at Medinet Habu. The inscriptions tell of how these various peoples overran the great powers of the day – the Hittites, the Mycenaeans, the Canaanites, the Cypriots and others – until they were finally stopped in Egypt, first by the pharaoh Merneptah in 1207 BC, and then again by Ramesses III himself in 1177 BC. However, the identity of the Sea Peoples, and their lands of origin, have for decades proved to be almost as elusive as the sea-god Poseidon, whose anger the Greeks blamed for earthquakes and what we would now call tsunamis. A century and a half after the first use of the term 'Sea Peoples' by an archaeologist, there is still no incontrovertible archaeological evidence for where they came from, who they were or what lands they conquered. The Shardana have been speculatively linked with Sardinia, the Peleset with the Philistines, who came from Crete, according to the Bible, and settled in Palestine; but there is no proof of these associations. Most probably, the Sea Peoples never really existed as a group.

The earthquake in about 464 BC that struck Sparta (which is located in the Peloponnese like Mycenae and Pylos), is one of the very few earthquakes of antiquity that was fairly reliably reported. Thucydides, writing his *History of the Peloponnesian War* in the same century, noted that it led to a revolt against Sparta by its subject peoples, the Helots of Laconia and Messenia. A later Greek historian, Diodorus Siculus, agreed, calling the earthquake 'a great and incredible catastrophe', which was part of a history of earthquakes in the area. (Diodorus may have been thinking of a major seismic predecessor during the 6th century, supposedly predicted by the natural philosopher Anaximander according to Cicero, for which there is no other evidence.) Today, a fault scarp 10–12 metres (33–39 feet) high and about

20 kilometres (12 miles) in length passes within only a few kilometres of the site of ancient Sparta. Diodorus is the source of the claim that more than 20,000 Spartan citizens died in the earthquake under collapsing walls, as a result of the 'tumbling down of the city and the falling of the houses' over a long period. Such a high number of Spartan casualties is thought by modern historians to be an exaggeration. Nevertheless, the Spartans were in due course obliged to recruit non-citizens as hoplites for their army, who did not fully subscribe to the famously disciplined Spartan military code.

Without doubt, the earthquake was a factor in the origin of the Peloponnesian wars between Sparta and Athens, which began in 460 BC and lasted, with intervals, until the end of the 5th century. For when the subject Helots saw that a majority of the Spartans had perished, 'they held in contempt the survivors, who were few', writes Diodorus.[11] They took the opportunity of the earthquake to rebel against the Spartans, who requested help from the Athenians, who in response sent a force of approximately 4,000 Athenian hoplites to Sparta led by Cimon, a military hero from the wars with the Persians. However, their presence in Sparta disturbed the regime. According to Thucydides, the Spartan aristocracy feared that the Athenian soldiers would make common cause with the Helots. So the hoplites were sent back to Athens: a humiliation that permanently soured Sparta's relationship with Athens.

Although Sparta emerged victorious from the Peloponnesian wars in 404 BC, it soon began to decline, as compared with Athens, during the 4th century. Its citizens were now greatly outnumbered by non-citizens, as noted by Aristotle, falling from about 8,000 Spartiates in 480 BC to not many more than 1,000 in 371 BC. By 250 BC, Sparta could muster only 700 hoplites to defend itself. To what extent this fall-off was due to the earthquake, rather than Sparta's loss of citizen manpower during its long struggles with Persia and with Athens, or to fatal flaws in the Spartan social structure, is debatable. However, 'there is no room for doubting the causal link between the massive Helot uprising of circa 464, with all that it implied for Spartan foreign and domestic policy, and the immediately preceding earthquake', writes a leading historian of Sparta, Paul Cartledge. 'The doubts concern rather the size of the quake . . . and its longer-run effects.'[12]

By an odd coincidence, the date of the earliest reported earthquake in ancient Rome is almost the same as that in ancient Greece. In 461 BC, according to the historian Livy, writing in the final decades of the 1st century BC, 'The ground was shaken by a violent earthquake.'[13] Another, in 83 BC, damaged Rome's public buildings and houses, and was treated as an omen of civil war. Throughout its history, up to the present day, the city and its rulers have suffered from both large and small earthquakes, 'and the chronicle of these events is one of the best in the world because of Rome's long historical record', write three scientists in *The Seven Hills of Rome: A Geological Tour of the Eternal City*.[14] Two earthquakes in AD 443 and 484 damaged the Colosseum, as is clear from a tablet near the entrance to the building thanking Decius Marius Venantius Basilius for his generosity in subsidizing its repair; and another major earthquake in 1349 most likely damaged the Colosseum beyond repair and left the city 'prostrate', according to Petrarch, the poet, who visited Rome in 1350.[15] The 1349 quake was also probably responsible for rotating some of the 30-ton blocks of Carrara marble stacked on top of each other to create the 42-metre-high (138 feet) column of Marcus Aurelius in the 2nd century AD. Their dislocation by about 10 centimetres (4 inches) decapitated a winged Victory and several Roman soldiers in the bas-relief carved on the column, while leaving intact the nearby column of Trajan, which stands on firmer foundations – as is still visible. '[S]ome of the great rents in the Colosseum, and . . . the fractures of the marble shafts of the columns of the Forum . . . prove that they were overthrown, not by Gothic hands, but by earthquake impulse', observed the pioneering seismologist Robert Mallet during a visit to Rome after he had measured the catastrophic effects of a great earthquake around Naples in 1857.[16] In 2000, in honour of the new millennium, Rome's monuments were at long last retrofitted to withstand future earthquakes.

That said, no earthquake in Rome – unlike in Sparta – was ever severe enough to shake the stability of the Roman state. Other parts of Italy suffered from much greater shaking than the capital city, as they do today. For example, an earthquake rocked Naples in AD 64, while the Emperor Nero was performing in the main theatre, according to the historian Suetonius. As soon as the spectators left, 'the theatre collapsed without harming anyone', notes the *Annals* of Tacitus.[17] A Naples inscription from around the same period, recording repairs to several public buildings, refers to earthquakes in the Latin plural, as '*terrae motibus*'. Pliny the Younger, in his famous letter describing the volcanic eruption of Vesuvius near Naples that devastated Pompeii and Herculaneum in AD 79, states that earth tremors preceding the eruption did not alert or alarm the populace, 'because they are frequent in Campania'.[18]

Indeed, only a few years before this catastrophic event, Pompeii and Herculaneum had come through a severe earthquake, either in 62 (according to Tacitus) or in 63 (according to Seneca, the tutor of Nero). Against Nero's advice to abandon Pompeii, much of the city was rebuilt in the 60s and 70s, initially without financial help from Rome. There is no firm historical evidence for a drop in the city's population, and no contemporary suggestion, whether from Seneca or others, that the earthquake might presage trouble from Mount Vesuvius. 'If some people fled, it was because of the threat of earthquakes, not because of a perceived threat from Vesuvius', notes a recent British Museum publication on Pompeii and Herculaneum.[19]

In Pompeii, for instance, the excavated house of a banker, Lucius Caecilius Jucundus, contains a substantial memorial to the AD 62/63 earthquake: a shrine with two marble reliefs on either side. One sculpted scene shows the temple of Jupiter in the forum and a neighbouring building dramatically tilting over towards the left, next to an upright scene of animal sacrifice by a priestess (presumably performed after the earthquake). The other depicts two yoked donkeys galloping for safety away from the collapsing Vesuvius Gate near the city's aqueduct. In Herculaneum, an inscription dated AD 75 records how the Emperor Vespasian restored the temple of Magna Mater, the great mother goddess Cybele, after its collapse in the 62/63 earthquake ('*terrae motu conlapsum*'). Another record reveals that Vespasian sent a commissioner to Pompeii to sort out various abuses, such as the illegal occupation of land after the earthquake. Overall, the predominant impression is not of a city in decline. Many buildings of the forum of Pompeii were fully repaired and newly decorated by the time of their second destruction by Vesuvius in 79.

Compared with the Italian peninsula, the Roman empire in the east experienced greater earthquakes, especially in Turkey, where, as we know, seismic activity probably contributed heavily to the collapse of the Bronze Age civilizations at sites such as Troy, Miletus and Hattusas. Consider the history of Antioch during classical antiquity. Had Antioch not suffered from multiple major earthquakes, the city might still be a living embodiment of its classical past like Rome.

Founded by one of Alexander the Great's former generals, Seleucus I, in 300 BC, Antioch fell under Roman rule in the 1st century BC, and became the capital of the Roman province of Syria in 64 BC. Julius Caesar, passing through Antioch during Rome's civil war, endowed it with splendid buildings, including a circus, later added to by Augustus, Herod and Tiberius. Mark Antony and Cleopatra got married there (almost certainly) in 37/36 BC. By the mid-4th century AD, claimed Edward Gibbon in his *Decline and Fall of the Roman Empire*:

> Fashion was the only law, pleasure the only pursuit, and the splendour of dress and furniture was the only distinction of the citizens of Antioch. The arts of luxury were honoured, the serious and manly virtues were the subject of ridicule, and the contempt for female modesty and reverent age announced the universal corruption of the capital of the East.[20]

When in AD 362–3 the ascetic, pagan, Roman emperor known as Julian the Apostate stayed in Antioch for nine months before launching his fateful campaign against the Persians (in which he died), he alienated the entire population, ranging from the luxury-loving city fathers uninterested in religious rituals to the land-hungry poor angered by his gifts of public land to rich men. The city's Christian population naturally despised Julian. Not only had he abandoned his imperial predecessor Constantine's Christian faith on becoming emperor in 360, he had also rejected Antioch's glory as the first place in which the disciples of Christ were called Christians, during the 1st century AD. There Saint Paul and Saint Barnabas preached the gospel in the streets, Saint Paul planned his missionary journeys, Saint Matthew supposedly wrote his gospel and Saint Peter was regarded as the city's first bishop. Soon after their time, Bishop Ignatius was martyred by being taken all the way from Antioch to Rome and fed alive to wild beasts in the Colosseum, according to Christian tradition; his remains were carried back to Antioch by his companions and buried outside the city's gates.

Antioch's first recorded earthquake was in 148 BC, followed by another in 130 BC. In AD 115 – when the Roman empire reached its greatest extent – there was a severe earthquake in Antioch. Many died, including one of Rome's consuls, during a visit by the Roman emperor Trajan, who was injured and spent the rest of his visit living in the open in the circus. In 458, nearly all of the buildings in Antioch were destroyed by an earthquake. The city was rebuilt; but another earthquake in 526, followed by a year and a half of aftershocks, culminating in a second earthquake in 528, proved catastrophic. About 250,000 people were killed in the 526 earthquake and fire – many of them probably visitors from the surrounding countryside present to celebrate Ascension Day – and the octagonal Great Church, built by Constantine after his conversion to Christianity, was destroyed. In Constantinople, the Byzantine (East Roman) emperor Justin I publicly lamented the ruination of Antioch. But he and his successor, Justinian the Great, nevertheless strove to rebuild both the church and the city, which was renamed Theopolis ('City of God').

In the end, it was not the earthquakes of the 520s that led to the abandonment of Antioch. Faction fighting and rioting (centred on the chariot races at the circus), religious strife, the outbreak of bubonic plague and the sacking of the city by the Persians in 540, 573 and 610, together ended its glory. By the time of Antioch's capture by Arab forces in 637, the city had been reduced to little more than a frontier fortress, which the Arabs renamed Antakya.

The abandonment, possibly around the same time, of another famous Near Eastern city, 'rose-red' Petra, in Jordan, also appears to have involved a massive earthquake or earthquakes. Petra, a centre of the spice trade with China, Egypt, Greece and India, was the capital of the Nabataeans, an Arab tribe who ruled the surrounding region from the early 3rd century BC, after the time of Alexander, until AD 106. Then Nabataea was annexed by Trajan into the Roman province of Arabia, in which Petra continued to flourish until changing trade routes brought about its gradual commercial decline. Most of its celebrated buildings date from the pre-Roman, Nabataean period. Working in the local sandstone with the help of craftsmen from Alexandria, the Nabataeans carved the elaborate, classically influenced, rock-cut buildings today known as the Treasury and the Monastery, a colonnaded street and many other buildings, some of which remain standing, while others lie in remarkably preserved ruined form. Analysis of these ruins, especially the fallen columns, suggests that they collapsed in a major earthquake. This probably occurred in AD 363 – the date of an earthquake known to have damaged neighbouring Jerusalem – according to the evidence of coins discovered in a belt purse belonging to a crushed female skeleton in one of the city's Roman houses; the coins were minted after a Roman currency reform in 354. But it is far from clear how this 363 earthquake affected the habitability of Petra. A second, post-Roman-period earthquake in 551, centred in Lebanon, may have further damaged the city, before Petra, like Antioch, was finally occupied by the forces of Islam in the 7th century, and eventually forgotten in Europe until the discovery of its amazing ruins by a Swiss explorer in 1812.

By the time of Petra's decline, China had embarked on the earliest attempt to measure earthquakes, setting the tone for the country's unique relationship with seismicity. Although the Chinese have long suffered from major earthquakes, including some of the most deadly in world history, they did not develop any earthquake mythology – in striking contrast to the Japanese. From the beginning of the relationship, the Chinese intelligentsia took a practical attitude towards earthquakes, seeing them as fundamentally natural, not divine, phenomena. China boasts not only the oldest record of an earthquake (780 BC) but also the oldest instrument to measure an earthquake.

The world's first seismometer was invented in AD 132 by a Chinese astronomer and mathematician at the imperial court, Zhang Heng, also known as Choko and Tyoko (modifications of the Japanese form of his Chinese name). His device consisted of eight hollow dragon-heads facing the eight principal directions of the compass. They were mounted on the outside of an ornamented vessel said to resemble a wine jar approximately 2 metres (6.5 feet) in diameter. Around the vessel's base, directly beneath the dragon-heads, were eight squatting toads with open mouths. In the event of an earthquake, a bronze ball would drop from a dragon-head into a toad's mouth with a resonant clang; the direction of the earthquake was supposedly indicated by which dragon-head dropped its ball, unless more than one ball dropped, indicating a more complex seismic shaking. The mechanism inside the seismometer is unknown. Seismologists of the 19th and 20th centuries therefore speculated about it, and several built working models. Whatever its precise arrangement, the mechanism must have comprised a pendulum of some kind as the primary sensing element, somehow connected to lever devices that caused the bronze balls to drop. That said, it is difficult, if not impossible, to explain how such an arrangement could have deconstructed a seismic vibration sufficiently to determine the true direction of a distant earthquake.

Nonetheless, according to a Chinese history, *Gokanjo* ('History of the Later Han'), in AD 138 Zhang Heng's seismometer is said to have enabled him to announce the occurrence of a major earthquake at Rosei, 650 kilometres (400 miles) to the northwest of the Chinese capital, Loyang, two or three days before news of the devastation arrived via messengers on galloping horses. His prediction apparently restored the faith of those who had doubted the exotic instrument's efficacy, and led the imperial government to appoint a secretary to monitor the behaviour of the seismometer, which remained in existence for four centuries.

This scientific Chinese attitude to earthquakes seems to have carried over into effective official disaster response. Unlike some other natural disasters, such as floods, great earthquakes never threatened the existence of an imperial Chinese government over two or three millennia. This was true even after an earthquake in 1556, during the Ming dynasty, with an estimated magnitude of 7.9, claimed up to 830,000 lives in Shaanxi province and nine neighbouring provinces: a world record for fatalities in a single earthquake. The victims were mainly farmers and their families living in elaborate caves traditionally known as *yaodongs* ('house caves') carved out of the dust blown from the Gobi desert, known as 'loess', that covers much of central China; millions of Chinese still live in seismically vulnerable *yaodongs*. (From 1936 to 1948, *yaodongs* in Shaanxi provided shelters for Mao Zedong and his

Communist revolutionaries.) Not only did these caves collapse, they were also buried in landslides, as the landscape was radically transformed to produce new hills, valleys and streams. 'The heavens crack, and the earth shakes' is one of many Chinese sayings about high politics and everyday life.[21] But not until after the foundation of the People's Republic of China in 1949 did the saying demonstrate its relevance after a great earthquake: the Tangshan earthquake in 1976, which quickly proved influential in the ending of the Maoist regime.

Japan's earthquakes have had more effect on Japanese history than China's have on Chinese history, despite costing fewer lives – probably because of Japan's far smaller area. The earliest reliable Japanese earthquake report (AD 599) occurs in the *Nihon Shoki* chronicle, compiled in AD 720. But the earliest actual listing of Japanese earthquakes is only a little more than a millennium old. Dated AD 900, it documents 700 earthquakes before AD 887. As for earthquake archaeology in Japan, unlike in Europe and the Middle East, it is severely constrained by the fact that old Japanese buildings have rarely survived in the archaeological record because of their perishable methods of construction, based on a pounded-earth foundation platform with wattle-and-daub walls and wooden load-bearing posts for the roof. Bricks, whether fired or unfired, were not used in Japan until the 19th century.

During this traditional period, the most serious earthquakes to ravage Edo (Tokyo) were those of 1703 and 1855. The first of these claimed around 2,300 lives, and an estimated 100,000 further lives in the tsunami produced by the earthquake. The second, known as the Ansei earthquake, provoked the popular outpouring of catfish prints (*namazu-e*) mentioned earlier. Although its magnitude has been estimated as comparatively low, between 6.9 and 7.1, its shallow focus and epicentre near the heart of the city caused substantial fatalities and property damage, mainly from fire: between 7,000 and 10,000 people died in and around Edo, and at least 14,000 structures were destroyed. Multiple aftershocks, as many as eighty per day, continued for nine days after the earthquake.

However, in the long run, the psychological effects of the Ansei earthquake of 1855 were more important than its physical ones. It struck the capital at a time of political stagnation in Japan's ruling class, the late Tokugawa shogunate, and one that was ominously close to the two visits in 1853 and 1854 of the United States naval officer Commodore Matthew Perry and his steamship squadron: a famous example of gunboat diplomacy, which began the opening of isolationist Japan to western trade and influence. Certain of the catfish prints of 1855 even link the earthquake directly to Perry's visit in their images and captions, encapsulating the ambivalent mixture of rejection and fascination in the Japanese reaction towards the American intruders.

In one print, the *namazu* has morphed into a threatening black whale spouting coins – not from its blowhole but rather from the place where a smoke stack is located on a steamship. Thus, the whale is intended to resemble one of Perry's 'black ships'. Japanese people standing on the shore beckon the wealth-creating whale-*namazu* to come closer. In another woodblock print, two figures – a kneeling, anthropomorphized, *namazu* clad in a kimono (with a builder's trowel near its feet and fishy tail) and Commodore Perry (with a rifle near his feet) – literally engage in a neck-to-neck tug-of-war, judged by a Japanese referee. Victory is going to neither side, but the catfish seems to have the edge over the US officer, who has been dragged slightly forward as the referee hails the fish. An extensive accompanying text contains a dialogue between Perry and the *namazu* that contrasts an aggressive America, effectively governed, with a Japan whose ineffective, feudal, Tokugawa government means that its people must turn for help to benevolent deities (such as Kashima, the controller of the *namazu*). Indeed, the earthquake-causing *namazu* is presented as partly beneficial to Japan. 'For Edo residents, the earthquake of 1855 was an act of *yonaoshi*, or "world rectification"', writes Gregory Smits, a historian of Japanese earthquakes. 'In this view, the Ansei earthquake literally shook up a society that had grown complacent, imbalanced, and sick.'[22] Although it would be too much to say that the Ansei earthquake was principally responsible for the social discontent and subsequent modernizing movement in Japan that eventually overthrew the Tokugawa shogunate and led to the restoration of the Meiji emperor in 1868, the earthquake undoubtedly played a major role, particularly through the subversive messages contained in the *namazu-e*. 'The anonymous print makers of Edo posited that the earthquake under their city had shaken up all of Japan, and they were right', concludes Smits.[23]

After 1868, Japan's modernizing movement eagerly embraced western science and expatriate western scientists, including the emerging discipline of seismology. Within a few decades, the Japanese had evolved a vigorous, indigenous, school of seismology. By the time of the Great Kanto earthquake in Tokyo in 1923, the earthquake mythology of the *namazu* was no longer relevant. But to understand these developments, we must first return to Europe, and the influence of certain European earthquakes – first those in Britain in 1750, and then one in Portugal in 1755 – upon the 18th-century Age of Enlightenment.

CHAPTER

2 THE YEAR OF EARTHQUAKES: LONDON, 1750

Earthquake alarm in central London, engraving, 1750. It is 'addressed to the Foolish and Guilty, who timidly withdrew themselves on the Alarm of another Earthquake'.

During 1750, there were five notable earthquakes in Britain. Two of them, on 8 February and 8 March, struck London and the southeast; the third, in mid-March, struck Portsmouth and the Isle of Wight on the south coast; the fourth, in early April, struck the northwest of England and the northeast of Wales; and the fifth, in late September, struck central England, in Northamptonshire and the surrounding counties. So much shaking in a single year was unprecedented in Britain.

In December, the Reverend William Stukeley – a fellow of the Royal Society now probably best known as the original source for the classic story of the falling apple and the theory of gravity told to him by his friend, Sir Isaac Newton – addressed the assembled fellows in London on 'The Philosophy of Earthquakes'. Stukeley proclaimed:

> The year 1750 may . . . be called the Year of Earthquakes . . . For, since they began with us at London, as far as I can learn, they have appeared in many parts of Europe, Asia, Africa, and America, and have likewise revisited many counties in our island: at length, on 30th of last September, taken their leave (as we hope) with much the most extensive shock we have seen in our days.[1]

In truth, 1750 was not at all a remarkable year for earthquakes in the world as a whole; Stukeley was misinformed. But he was surely right about Britain, and his proposed label, the 'Year of Earthquakes', has stuck.

Although the first of these British earthquakes, soon after half past midday on 8 February, was small – with an estimated magnitude of just 2.6 according to today's British Geological Survey – its epicentre was shallow and centred beneath the capital, apparently around London Bridge. So the city received a considerable jolt. The lord chancellor was then sitting in Westminster Hall with the Courts of King's Bench and Chancery and others. For a moment everyone thought that the great edifice was going to collapse on their heads. In Lincoln's Inn Fields, Newcastle House trembled so much that the duke of Newcastle sent out his servant to enquire what had happened from a neighbour, the physicist Gowin Knight (later, first principal librarian of the British Museum). The servant found Knight busy investigating the signs of disturbance in his own residence: a grate that had been observed to move, a fire-shovel that had been knocked over and even a bed that had moved from its proper position. In Gray's Inn, a lamp-lighter very nearly fell from his ladder. At Leicester House, home of the prince of Wales, the foundations were believed to be sinking. Throughout the City and Westminster people felt their desks lurch; chairs shook; doors slammed; windows rattled; and pewter and crockery clattered on its shelves. In Leadenhall Street, the location of the East India Company, part of a chimney fell. In Southwark, south of the River Thames, a slaughter-house with a hay-loft collapsed. The intensity of the shaking seems to have been greatest in the areas bordering the river.

To begin with, an earthquake was not accepted as the explanation – so improbable did an earthquake in London appear to be. The last real shock in London had occurred in 1692, probably as a distant aftershock of the great earthquake that destroyed Port Royal in the Caribbean, with its centre across the English Channel in the Low Countries. Having an estimated magnitude of 5.7, it had induced fear and panic in London, and thronged the streets with confused crowds. But by 1750 this event was too long ago to be remembered. Instead, there were theories about cannon-fire and exploding gunpowder magazines. Then, after the idea of a London earthquake became unavoidable, it was said that Newton, before his death in 1727, had predicted the jolt by calculating that the planet Jupiter would approach close to the earth in 1750. 'You know we have had an earthquake', Horace Walpole, man of letters and member of Parliament, wrote to his friend, Sir Horace Mann, near the end of February. 'I am told that Sir Isaac Newton foretold a great alteration in our climate in the year '50, and that he wished he could live to see it. Jupiter, I think, has jogged us three degrees nearer to the sun.'[2] But this was almost all that Walpole had to say about the earthquake. Clearly, after two or three weeks, people were beginning to forget the strange experience.

Then, exactly four weeks after the first shock, at 5.30 a.m. on 8 March, came the second. It was more pronounced than the first (estimated magnitude 3.1) and more widespread, covering five times the area: an approximate circle of diameter 63 kilometres (39 miles) with its centre about 3–5 kilometres (2–3 miles) north of London Bridge. Two houses in

Whitechapel collapsed, and several chimneys fell in various parts of London, as did stones from the new towers of Westminster Abbey.

Walpole was in bed at his house in central London at the time. Three days later, he reported at some length to the same friend:

> on a sudden I felt my bolster lift up my head; I thought somebody was getting from under my bed, but soon found it was a strong earthquake, that lasted near half a minute, with a violent vibration and great roaring. I rang my bell; my servant came in, frightened out of his senses: in an instant we heard all the windows in the neighbourhood flung up. I got up and found people running into the streets, but saw no mischief done: there has been some; two old houses flung down, several chimneys, and much china-ware. The bells rung in several houses.[3]

In fact, the mean duration of the shock is estimated to have been a mere 5.4 seconds. The president of the Royal Society, antiquarian Martin Folkes, was also in bed in central London at the time. Reporting to the society that very day, he noted that the vibration and noise could not have been that of a passing cart or coach – to which many compared it – because everything was entirely quiet at such an early hour. He also noted that the shock had been felt on the outskirts of London, to the north:

> I sent a servant out about 7 o'clock, and he met a countryman, who was bringing a load of hay from beyond Highgate, and who was on the other side of the town when the shock happened; he did not, he said, feel it, as he was driving his waggon; but that the people he saw in the town of Highgate were all greatly surprised, saying they had had their houses very much shocked, and that the chairs in some were thrown about in their rooms.[4]

Indeed, near Holland House, in the western part of London, the bailiff of the secretary of war, Henry Fox (later the first Lord Holland), who was counting his sheep, observed the dry, solid ground move like a quagmire or quicksand, causing much alarm among the animals and some crows nesting in nearby trees. In and around London, 'cats started up, dogs howled, sheep ran about, a horse refused to drink, the water being so much agitated, in several ponds fish leaped out of the water and were seen to dart away in all directions', notes Charles Davison in *A History of British Earthquakes*.[5]

This time around, there was genuine alarm among the city's human inhabitants, too. Walpole, who did not take the threat of a major earthquake seriously, observed people's reactions over the next weeks with a mixture of irony and contempt. 'Several people are going out of town, for it has nowhere reached above ten miles from London: they say, they are not frightened, but that it is such fine weather, "Lord! one can't help going into the country!"' he mentioned on 11 March.[6]

A slight tremor occurred on 9 March, and then came a powerful rumour that there would be a third shock exactly a month after the second one, on 7–8 April, 'which is to swallow up London', Walpole wryly commented on 2 April.[7] The rumour was started by an army trooper who was not in his right mind and would eventually be despatched to Bedlam, London's lunatic asylum. By 4 April, doomsday had somehow advanced to the very next day, and panic took hold. 'This frantic terror prevails so much, that within these three days seven hundred and thirty coaches have been counted passing Hyde Park corner, with whole parties removing into the country', Walpole reported from the frontline. 'Here is a good advertisement which I cut out of the papers today: "On Monday next will be published (price 6*d.*) A true and exact list of all the nobility and gentry who have left, or shall leave, this place through fear of another earthquake."' He continued vividly:

> Several women have made earthquake gowns; that is, warm gowns to sit out of doors all tonight. These are of the more courageous. One woman, still more heroic, is come to town on purpose: she says, all her friends are in London, and she will not survive them. But what will you think of Lady Catherine Pelham, Lady Frances Arundel, and Lord and Lady Galway, who go this evening to an inn ten miles out of town, where they are to play at brag [an ancestor of poker] till five in the morning, and then come back – I suppose, to look for the bones of their husbands and families under the rubbish.[8]

That Walpole was not exaggerating is confirmed by the 'Historical Chronicle' of April published shortly after the events in the monthly *Gentleman's Magazine*. For 4 April, this reads:

> Incredible numbers of people, being under strong apprehensions that London and Westminster would be visited with another and more fatal earthquake, on this night, according to the predictions of a crazy lifeguardsman, and because it would be just 4 weeks from the last shock, as that was from the first, left their houses, and walked in the fields, or lay in boats all night; many people of fashion in the neighbouring villages sat in their coaches till daybreak; others went to a greater distance, so that the roads were never more thronged, and lodgings were hardly to be procured at Windsor; so far, and even to their wit's end, had their superstitious fears, or their guilty conscience, driven them.[9]

Perhaps needless to add, the earth did not move on the supposed day of doom, 5 April. However, in June there was a 'loud report like that of a cannon' at London and Norwich, without any tremor.[10]

Part of the blame for the panic must undoubtedly fall on religious preachers: 'the clergy, who have had no windfalls of a long season', commented Walpole.[11] On 9 March, the day after the second earthquake, Charles Wesley, one of the founders of the growing Methodist movement, gave his sermon 129, 'The Cause and Cure of Earthquakes', in which he informed Londoners: 'God is himself the Author, and sin is the moral cause.'[12] At the same time, he composed a hymn, with these first and last verses:

> Woe to the men on earth who dwell,
> Nor dread the Almighty's frown;
> When God doth all His wrath reveal,
> And shower His judgements down.
>
> Firm in the all-destroying shock
> [We] may view the final scene;
> For lo! the everlasting Rock
> Is cleft to take us in.[13]

In 1777, after an earthquake had alarmed the city of Manchester, John Wesley (brother of Charles), no doubt recalling what had happened in London in 1750, remarked to a friend: 'there is no divine visitation which is likely to have so general an influence upon sinners as an earthquake'.[14] The only cure he recommended was, of course, repentance.

At the same time, a leading clergyman, William Whiston, who had been the successor to Newton as Lucasian professor of mathematics at Cambridge University, took the opportunity of dilating on his personal long-held belief that the end of the world was quite close at hand, as predicted by ninety-nine signals. Whiston's signal No. 92 was that there would occur a terrible – but to good men a joyful – earthquake, which would destroy one-tenth of an eminent city. The shock to London on 8 February was too tempting for Whiston to resist, and he gave three vehement lectures in the city: the first one covering the ninety-nine signals, the second on their fulfilment and the third attacking the wickedness of the present age. The second lecture, on 8 March, happened to coincide with the second shock. None of the three was well attended. A fellow clergyman and future bishop, William Warburton, then a preacher to Lincoln's Inn, joked in a private letter that: 'The greatest mischief the earthquakes have hitherto done is only widening the crack in old Will Whiston's noddle'.[15] Yet, Whiston's ideas were talked about, given his reputation as a theologian, biblical historian, astronomer, mathematician and physicist; and they certainly contributed to the March malaise in London.

Another scientifically minded divine and fellow of the Royal Society, the Reverend Roger Pickering, pastor of a London church of Dissenters, took a more measured view of the earthquake than Whiston. He argued against panic and counselled his flock to remain in London. His sermon on 1 April took as its theme the omnipresence of God. According to the psalmist, Pickering reminded his listeners, 'If I take the wings of the morning, and dwell in the uttermost parts of the sea: even there shall thy hand lead me, and thy right hand shall hold me up.'[16] In other words, said the reverend, no matter how far a person might travel or at what velocity, no one could escape divine vengeance if they were guilty, nor be beyond the protection of God if they were righteous – even if they were able to travel to the furthest point possible, half-way around the globe, at the speed of light in, say, one-thousandth of a minute. Therefore, what possible good could come of travelling a few miles out of London, in order to evade an earthquake? Instead, Christians must be courageous and trust in God.

But it was unquestionably the warnings of the bishop of London, Thomas Sherlock, a former vice-chancellor of Cambridge University, which attracted the most attention during March and early April 1750. Sherlock's *A Letter to the Clergy and People of London and Westminster . . . on Occasion of the Late Earthquakes*, published on 16 March, sold 10,000 copies in two days, according to Walpole; was reprinted several times; and is said to have sold more than 100,000 copies in less than six months, 40,000 of which were purchased by subscription for free distribution among the poor.

The following extracts from the *Letter* in the *Gentleman's Magazine* give a good idea of its tone. The bishop opens:

> It is my duty to call upon you, to give attention to all the warnings which God in his mercy affords to a sinful people: such warning we have had, by two great shocks of an earthquake; a warning, which seems immediately directed to these great cities, and the neighbourhood of them; where the violence of the earthquake was so sensible, tho' in distant parts hardly felt, that it will be blindness wilful and inexcusable not to apply to ourselves this strong summons, from God, to repentance.

Then Sherlock takes a crack at those natural philosophers (the 18th-century term for scientists), such as certain fellows of the Royal Society, who allow insufficient space to God in their explanations:

> Thoughtless and hardened sinners may be deaf to these calls; and little philosophers, who see a little, and but very little into natural causes, may think they see enough to account for what happens, without calling in the aid of a special providence, not considering that God who made all things, never put anything out of his own power . . .[17]

Warburton approved of Sherlock's letter, unlike Whiston's sermons, calling it a 'very primitive discourse, and what is more, a very good one'.[18]

But the sceptical Walpole, who had expected more sense from Sherlock, was incensed, writing to his friend Mann:

> You never read so impudent, so absurd a piece! This earthquake, which has done no hurt, in a country where no earthquake ever did any, is sent, according to the Bishop, to punish bawdy prints, bawdy books . . . gaming, drinking – (no, I think, drinking and avarice, those orthodox vices are omitted), and all other sins, natural or not.[19]

Walpole's friend, Richard Bentley, of like mind, responded to Sherlock with an anonymous pamphlet of his own, in the form of a satirical letter supposedly describing the much-predicted third earthquake in London. It was summarized by T. D. Kendrick, a 20th-century director of the British Museum, in the London earthquakes preamble to his study of the great Lisbon earthquake of 1755:

> When [the earthquake] happened the first man sunk was the bishop of London, though he might have escaped if he had not been so busy distributing copies of his letters. The duke of Newcastle was next, the place of his disappearance being marked by scatterings of papers and red tape. Then followed a long list of other notable casualties, and the news that Mr Whiston had set out on foot for Dover on the way to Jerusalem to meet the millennium.[20]

The fiasco in early April 1750 created a general air of sheepishness in London society, especially among those who had decamped from the capital in fear for their lives. But within a matter of months, the London earthquakes and their embarrassments began to be forgotten by Londoners. Natural philosophers throughout the country, however, became fascinated by the subject. During 1750, the 'Year of Earthquakes' provoked many fellows of the Royal Society to begin thinking about the causes of seismicity. By the end of the year, almost fifty articles and letters on the subject had been read before the society, which were promptly published as an appendix to its *Philosophical Transactions*. Although the majority of these were 'merely personal narratives', notes Davison, they have provided invaluable data for historians assessing the true magnitude, extent and influence of the earthquakes.[21] The serious study of earthquakes may be said to date from these Royal Society accounts.

After the February shock, an article in the *Gentleman's Magazine* commented, accurately enough, that there was as yet very little understanding of earthquakes:

> With respect to the cause of the dreadful phenomena naturalists are greatly divided; some supposing it to be water, others fire, and others air: but it is to be supposed, as the basis of each hypothesis, that the earth is full of subterraneous caverns, particularly about the roots of mountains, some of which are filled with water, others with exhalations, and that some parts of the earth are replete with nitre, sulphur, bitumen, vitriol, etc.[22]

In truth, despite the scientific revolution of the 17th century and the achievements of Newton in understanding the solar system, the science of the earth, including earthquakes and volcanoes, had advanced no further by 1750 than the musings of the ancient Greeks in the time of Aristotle. Some writers even took their cue from Aristotle's analogy between the caverns causing earthquakes and the pent-up winds inside human beings, and went so far as to anthropomorphize the earth's surface as an elderly face with 'warts, furrows, wrinckles and holes of her skin, which age and distempers have produced' – with the implication that earthquakes were like 'universal shivering fits' as if in an 'ague'.[23]

More respectably scientific theories included that of Newton's contemporary, the brilliant if erratic Robert Hooke. After noting in the 1660s the presence of fossil seashells on the tops of mountains, Hooke repeatedly tried to persuade the Royal Society that earthquakes had been responsible for raising up the world's great hills and mountains from the primeval ocean. But since both Hooke and the other fellows adhered firmly to the biblical view of the planet as being only a few thousand years old, they were compelled to postulate enormous seismic convulsions over a very short period, without offering any convincing mechanism. Eventually, Hooke proposed that the earthquakes that had caused the mountainous alterations in the earth's surface might have been triggered by a sudden displacement of the planet's axis of rotation, due to the fact that the earth was not spherical. But he could not persuade the astronomer royal, John Flamsteed, that the axis had actually varied since antiquity.

Flamsteed, for his part, came up with his own, quite different, theory of earthquakes, after experiencing the significant quake in London in 1692. In Flamsteed's radical view, earthquakes did not originate inside the earth, but instead in the air, where they were caused by the explosion of nitrous and sulphurous particles, which had escaped from underground. His argument was basically that there could be no caverns beneath England, suitable for the generation of earthquakes, because there appeared to be so little evidence of volcanic activity in the country (as compared with, say, earthquake-prone Italy).

Although neither Hooke nor Flamsteed themselves published their earthquake theories before their deaths (in 1703 and 1719, respectively), Flamsteed's was revived and published in 1750, along with a similar theory by another clergyman and natural philosopher, Stephen Hales. The latter favoured the idea of earthquakes as explosions of sulphurous fumes in the air, in a manner resembling that of lightning. As a result of the Flamsteed and Hales theories, 'The term "airquake" then enjoyed a brief currency in the popular press', comments historian of science Frances Willmoth.[24]

Hales's reference to lightning is significant. The new concept of electricity was coming into vogue in 1750. In the autumn of the previous year, the intriguing electrical experiments of 'Mr Franklyn of Philadelphia' (Benjamin Franklin), inventor of the lightning rod, had been communicated to the Royal Society. Now a brand-new earthquake theory, which would spark the imagination of both natural philosophers and the public, was based on electricity. In May 1750, the inimitable Walpole recorded, 'One Stukely, a parson, has accounted for [earthquakes], and I think prettily, by electricity – but that is the fashionable cause, and everything is resolved into electrical appearances, as formerly everything was accounted for by Descartes's vortices, and Sir Isaac's gravitation.'[25]

In Stukeley's above-quoted December address before the Royal Society at the end of the 'Year of Earthquakes', he laid out his reasoning with considerable eloquence and some logic, though without any convincing physical proof. He rejected the conventional, ancient Greek, explanation of fiery underground caverns causing earthquakes on the following grounds:

> We observe, the vulgar solution of subterraneous eruptions receives no countenance from all that was seen or felt during these earthquakes: it would be very hard to imagine how any such thing could so suddenly and instantaneously operate thro' this vast space, and that in so similar and tender a manner, over the whole, thro' so great a variety as well as extent of country, as to do no mischief.
>
> A philosophical inquirer in Northamptonshire, who had his eye particularly on this point, takes notice that there were not any fissures in the ground, any sulphureous smells, or eruptions, any where perceiv'd, so as to favour internal convulsions of the earth . . .
>
> The former earthquake, that happened at Grantham, Spalding, Stamford (which towns lie in a triangle) took up a space which may in gross be accounted a circle of 20 miles in diameter; the centre of which is that great morass called Deeping-Fen. This comprehends 14 miles of that 20 in diameter; and where, probably, the electrical impression was first made. Much the major part of Deeping-Fen is under water in the winter; underneath is a perfect bog: now it is very obvious how little favourable such ground is for subterraneous fires.[26]

As for his evidence in favour of an 'electrical impression', Stukeley adduced what appeared to be the instantaneousness of the seismic vibration over large distances:

> As far as we can possibly learn, where no one can be prepar'd at different places, by time-keepers, this mighty concussion was felt precisely at the same instant of time, being about half an hour after 12 at noon. This, I presume, cannot be accounted for by any natural power, but that of an electrical vibration; which, we know, acts instantaneously . . . which acknowledges no sensible transition of time, no bounds.[27]

That said, honesty compelled Stukeley to admit that he had absolutely no idea of the origin of the aforesaid electricity:

> How the atmosphere and earth are put into that electric and vibratory state, which prepares them to give or receive the snap, and the shock, which we call an earthquake, what it is that immediately produces it, we cannot say; any more than we can define what is the cause of magnetism, or of gravitation, or how muscular motion is perform'd, or a thousand other secrets in Nature.[28]

It was a lame conclusion and, unsurprisingly, Stukeley's theory faded away as the fashion for electrical explanations waned in the face of mid-18th-century scientific ignorance of electricity. Rather than Hooke's convulsions, Flamsteed's airquakes and Stukeley's electricity, Newton's mechanics eventually proved to be the most useful theory for understanding earthquakes – in the hands of yet another clergyman fellow of the Royal Society, John Michell, an astronomer at Cambridge University now best known as the first person to propose the existence of black holes in space (in 1783).

His attraction to earthquakes began with the events of 1750. As a direct result, Michell took up the challenge of analysing eyewitness reports and accounting for earthquake motions in Newtonian terms. But it was not until the Lisbon earthquake five years later that he could begin to produce the first theory of earthquakes that was a genuine advance on the views of the ancient Greek thinkers. In Britain, data collected in 1750 was now supplemented by data on the effects of the 1755 Lisbon earthquake gathered from all over Britain and the Continent by the Royal Society. In due course, Michell produced an important, if flawed, geological paper, 'Conjectures Concerning the Cause and Observations upon the Phaenomena of Earthquakes', which was published in the Royal Society's *Philosophical Transactions* for 1760.

Michell correctly concluded that earthquakes were 'waves set up by shifting masses of rock miles below the surface', although his explanation for this shifting relied wrongly on explosions of steam, as underground water encountered underground fires.[29] When the shifting occurred beneath the seabed, he also rightly concluded that it would produce a sea wave (a tsunami) as well as an earthquake. There were two types of earthquake wave,

Michell said, once again coming close to the truth: the first was a 'tremulous' vibration within the earth, followed shortly by an undulation of the earth's surface. From this he argued that the speed of an earthquake wave could be determined by its arrival times at different points on the surface. Such times were known approximately from eyewitness reports at far-flung places affected by the Lisbon earthquake, which enabled Michell to calculate a speed for its wave of some 1,930 kilometres (1,200 miles) per hour. He was the first scientist to attempt such a calculation – inaccurate though it was, and unaware though he was that the speed of seismic waves varies with the types of rock through which they pass. He then went further by theorizing that the surface origin of an earthquake, what we now call its epicentre, could be calculated by combining the arrival times of an earthquake's wave at many different locations. Although Michell curiously chose a different – and inaccurate – way to calculate the epicentre of the Lisbon earthquake, relying instead on reports of the direction of the tsunami, his theoretical principle for locating an epicentre is the basis of the method used today.

Despite being a clergyman, Michell left God out of his analysis. Such an omission was still far from the norm in natural philosophy during the 1750s. Stukeley, for all his love of electricity, nonetheless carefully included a reference to God (the 'Author of Nature') in his 1750 Royal Society lecture;[30] and so did most other scientific writers on earthquakes. In Walpole's laconic verdict, 'they all take care, after accounting for the earthquake systematically' – by whatever natural theory appealed to them – 'to assure you that still it was nothing less than a judgement.'[31]

The shadow of Bishop Sherlock's bestselling pastoral letter of 1750 was a long one. Five years later, even a scientifically minded clergyman like Warburton would find himself struggling to come to terms with the Lisbon catastrophe without invoking God. 'To suppose these desolations the scourge of heaven for human impieties, is a dreadful reflection', Warburton confided to a friend, 'and yet to suppose ourselves in a forlorn and fatherless world, is ten times a more frightful consideration.'[32] Unlike the genteel shaking of London in 1750, the cataclysmic earthquake in the Portuguese capital in 1755 would provoke a dramatic conflict between politics, religion, philosophy and science, with reverberations that are still disturbing our modern world.

CHAPTER

3 **THE WRATH OF GOD:** LISBON, 1755

Part of a mass grave in Lisbon containing victims of the earthquake in 1755, discovered in 2004 beneath what is now the city's Academy of Sciences.

It is no exaggeration to say that the sudden destruction of Lisbon on 1 November 1755 exerted an influence on European life and thought in the 18th century as far-reaching as the obliteration of Hiroshima and Nagasaki by atomic bombs in the 20th century. By the 19th century, images of a shaking Lisbon were icons of natural disaster comparable with the smothering of Pompeii and Herculaneum by the eruption of Vesuvius. The Lisbon earthquake had become 'a cultural shorthand for initiation into a sceptical, rational and self-consciously modern search for natural causes', notes a recent historian, Deborah Coen.[1]

Thus, in 1848, London staged a hugely successful exhibition telling the story of Lisbon's earthquake, tsunami and fire through the medium of 'moveable paintings' accompanied by dramatic music. The venue was the newly refurbished Colosseum theatre in Regent's Park. The *Illustrated London News* of 1850 described in rapt detail the experience of seeing this 'Cyclorama of Lisbon'. In the opening scenes, said the magazine,

> We are presented with the beautiful, varied, and sublime scenery of the [River] Tagus, the movement of which produces a peculiar feeling in the spectator. The theatre in which he sits seems like a vessel floating down the stream, and passing one object after another – the mountainous shore – the ships and vessels, the merchantmen and the *xebec* [Mediterranean warships] – the nunnery, the fort, the mansion, the palace, the various convents, the Consulate House, and, at length, the City, with its palatial, ecclesiastical, public and private buildings – all doomed to sudden destruction. The last scene presents the Grand Square of Lisbon, 'with its gorgeous palaces and magnificent ranges of streets, massive arches and noble flights of steps, vases, and other colossal decorations, with the beautiful statue and fountain of Apollo.'

The soundtrack consisted of excerpts from works such as Beethoven's *Pastoral Symphony*, Mozart's *Don Giovanni*, Mendelssohn's 'Wedding March' and Haydn's 'Il Terremoto' ('The Earthquake'), plus a Portuguese dance and a Brazilian melody – since it was Brazilian gold mined in Portugal's South American colony that had provided the funds to expand and embellish Lisbon in the half-century before its devastation. The music was blasted out by a grand organ with sixteen pedals and 2,407 pipes. When the earthquake struck, there was the sound of a 'subterranean roar', followed by an 'appalling crash', during which the stage was plunged into darkness.

After the lights came up again, the awful scene was reported by the magazine as follows:

> We next see the ships tossing upon the waves, fated to the destruction with which the lowering sky only too visibly threatens them. All is terror and despair. But this passes, and the site of the city returns, now covered with ruins where so lately we contemplated the glories of architectural genius – all, by the visitation of an inscrutable Providence, involved in one common wreck, with more than thirty thousand of its dwellers.

Overall, said the *Illustrated London News*, 'A more magnificent series of pictorial wonders cannot be imagined.'[2] So widely appreciated was the Colosseum's Lisbon earthquake 'cyclorama' that it continued into the 1850s. Indeed, it successfully competed against a display about the destruction of Pompeii at London's Great Exhibition in 1851.

Not long after the earthquake's centenary, Charles Dickens felt the allure of the disaster whilst on a visit to Portugal in 1858. In his journal, *Household Words*, Dickens conjured up a dreadful vision of an apocalypse in paradise brought about by the earthquake:

> That night, looking from the Bragança [Hotel] window at the weltering bay which seemed turned to silver, over which highway I could see away to Belém, the guarded mouth of the Tagus, I beheld the tranquil terraced roofs below, quiet in the moonlight; for the wilful Mohammedan moon was in her crescent, and I could almost imagine myself in the old Moorish city. As I looked, I fell into a reverie in my chair in the Bragança balcony. Napier's *Peninsular War* dropping from my hand, I imagined myself, that November morning, on that safe roof-top watching that tranquil city. Suddenly, the houses all around me began to roll and tremble like a stormy sea. Through an eclipse dimness I saw the buildings round my feet and far away on every side, gape and split; the

> floors fell with the shake of cannons. The groans and cries of a great battle were round me. I could hear the sea dashing on the quays, and rising to swallow what the earthquake had left. Through the air, dark with falling walls and beams, amid showers of stones red with the billows of fire from sudden conflagrations, I saw the cloudy streets strewn with the dead and dying; screaming crowds, running thickly, hither and thither, like sheep when the doors of the red slaughter-house are closed.[3]

Yet today, it has to be said the destruction of Lisbon is largely forgotten by the world – unlike the destruction of Pompeii. 'Nine out of ten well-educated, well-travelled Europeans are still unaware of it', admits Edward Paice in *Wrath of God*, his engaging account of the earthquake intended for a general readership, published in 2008.[4]

Even some academic specialists in the period neglect the disaster. 'It is a source of great puzzlement', comments a leading geographer, Peter Gould, 'that this devastating environmental event, right at the centre of the century of the Enlightenment, so often leaves hardly a trace in many standard works by biographers and historians.' If Voltaire had not written his celebrated poem on the disaster in 1755, followed by his story, *Candide*, in 1759, 'one wonders whether the event might have disappeared from human memory altogether', writes Gould.[5]

Perhaps this surprising amnesia can be explained, at least to some extent. Archaeologists, as we know, neglect earthquakes and other natural disasters as explanations of cultural change, preferring to look for human agency. On the whole, so do historians. That is why Hiroshima and Nagasaki will most likely never be forgotten as universal symbols of 20th-century disaster, whereas Tokyo and Yokohama have far less global resonance, despite their comparable, earthquake-related catastrophe. Pompeii – almost uniquely among natural disasters – is still remembered the world over partly because its residents intensified their predicament by refusing to heed the warnings from Vesuvius of its imminent eruption. The disaster at Pompeii was caused by both nature and man. The Lisbon earthquake, by contrast, struck without any foreshocks, giving no one in the city the slightest chance to escape its consequences.

Moreover, the ruins of Pompeii of course exist and have attracted millions of spectators, whereas the ruins of Lisbon – much visited in their day by tourists with a penchant for scenes of Gothic melancholy – were eventually cleared away in order to build a splendid new city. In addition, Portuguese writers avoided the earthquake as a subject for memoirs and secular literature, not only at the time but also in the decades thereafter. There is no Portuguese equivalent of Voltaire's poem about the earthquake. No memorable contemporary record of the earthquake's effect on Portuguese life was written, unlike, for example, the responses of major Japanese writers in the 1920s to the Great Kanto earthquake in Tokyo.

Indeed, almost everything we know about what actually happened in Lisbon on All Saints' Day in November 1755 and in the days that followed comes from the horrified accounts not of resident Portuguese but of affected foreigners, many of them resident British merchants pursuing a lucrative trade with Portugal. Portuguese witnesses of the earthquake, traumatized by events and operating under the watchful eyes of the court and government, the clergy and the Inquisition, were reluctant to express themselves on paper.

It is also the case that if Portugal had been a more influential country in 1755, then the Lisbon earthquake might well be better remembered today. But in fact, despite Lisbon's financial wealth and considerable size – as probably the fourth largest city in Europe after London, Paris and Naples – Portugal was regarded throughout Europe as an economic, political and intellectual backwater.

It manufactured almost nothing and imported almost everything, including its textiles, toys, watches, chandlery, arms and shot from England. These were paid for with gold from Minas Gerais in Brazil that had been discovered in the 1690s – to the tune of more than £25 million sent to Britain by the middle of the 18th century, thereby enabling London to become the world's foremost financial market. 'At the time of the earthquake Portugal was an aloof, proud, happy, spendthrift country,' writes T. D. Kendrick in *The Lisbon Earthquake*, 'forced to buy things that it should itself have produced, and slipping fast into financial dependence on London and Hamburg.'[6]

Portugal's ruler for the first half of the century, João V, spent his one-fifth share of these massive Brazilian revenues (the rate of the royal tax on the gold trade) on the construction of churches, monuments and palaces, such as the extraordinary palace-church-convent complex at Mafra, 40 kilometres (25 miles) north of Lisbon, on a plateau overlooking the Atlantic. 'When he wanted a festival, he ordered a religious parade. When he wanted a new building, he built a convent, when he wanted a mistress, he took a nun', wrote Voltaire.[7] The king's only practical construction was a great aqueduct; and Lisbon had no clubs and few entertainments on the model of London's, despite a lively street scene. João was reputed to be the richest monarch in Europe, and therefore rarely needed to assemble his court. The Portuguese parliament (the Cortes) was not called even once between 1698 and 1820. When João became ill in 1742, the affairs of state fell into the hands of the Catholic clergy: cardinals and priests, notably the Jesuit confessor of the king. The armed forces dwindled away, despite Portugal's extraordinary maritime history, and the king was effectively guarded by the church.

By 1750, the country had perhaps 200,000 clergy in a population of fewer than 3 million people, and was 'more priest ridden than any other country in the world, with the possible exception of Tibet', according to a leading historian of Portugal and the Far East, C. R. Boxer.[8] In that year – the 'Year of Earthquakes' in England – João V died, on the very same day as a slight earthquake in Lisbon. His son and successor, José I, was chiefly interested in riding, card playing, attending the theatre and opera, and worshipping God.

Although no foreshocks heralded the 1755 earthquake, it was preceded by some ominous natural phenomena in Lisbon and in other parts of the country. The weather was different from usual, as often seems to be the case before great earthquakes. The day before it struck, 31 October, was warm for the season, noted the consul of Hamburg, who was surprised to see a fog rolling in from the sea, as this was normally a phenomenon seen in the summer. Then the wind blew the fog back out to sea, where it became as thick as the consul could recall. Behind the receding fog, he thought he could hear the sea 'rise with a prodigious roaring'.[9]

All along the coast, the evening tide was late by two hours, causing wary fishermen to haul their boats further up the beach. Around the same time, a village fountain was observed to run almost dry. Elsewhere, a well altogether dried up. A physician noted that for several days there had been complaints from people about an odd taste in Lisbon's water supply. In another place, the air had a sulphurous smell.

Animals, too, behaved abnormally, which is again a common observation prior to great earth shocks, as well as during the time of earth shaking. Dogs, mules and caged birds became unaccountably agitated. Rabbits and other animals left their burrows. Worms crawled to the surface in large numbers.

But none of this was sufficient evidence to anticipate the imminence of an earthquake. (Indeed, the same is still true today, as we shall see in later chapters.) Lisbon had been struck hard once before, in 1531, when about 30,000 people had died. It had been severely shaken, too, though much less damagingly, in 1724, within living memory. And in 1750, it had received another shaking (on the day of the king's death). However, this most recent of earthquakes had been too slight to disturb anyone's complacency – unlike the earthquakes in London.

The violent shaking of 1755 began around 9.30 a.m. on 1 November. It lasted for between seven and ten minutes, as compared with about five seconds in London in March 1750 and four to five minutes in Tokyo in 1923. It came in three distinct waves separated by a pause of no more than a minute, of which the second wave was the greatest, with a magnitude later estimated as 8.5–8.8, on the evidence of the level of destruction. Within a mere quarter of an hour, the great city was 'laid in ruins', noted the English consul.[10] Tremors continued all through the day and night, with no more than a quarter of an hour's respite, culminating in the biggest aftershock since the first day one week later, on 8 November. During the next nine months there were said to have been 500 aftershocks. In 1761, there was a further major earthquake. It shook the city for at least three minutes, possibly as long as five minutes, during which many of the ruins of 1755 finally collapsed.

Since 1 November was All Saints' Day, many Lisboetas were attending Mass in the city's numerous and lavish churches when the earthquake began.

The coincidence was a horrible one for the worshippers, who were crushed under a rain of falling masonry, as many churches quickly crumbled. The palace of the Inquisition fell, too, as did the magnificent new opera house – its scheduled evening performance of *A Destruição de Troya* ('The Destruction of Troy') forever cancelled.

Other people were slaughtered in the conflagration started by kitchen fires. The blaze was not finally extinguished until almost a week after the earthquake. It had been so intense that the ruins were still too hot, seventeen days after the disaster, for merchants and others facing financial ruin to search for their valuables.

Yet others were drowned by a tsunami. Every Lisboeta who saw the wall of water rushing in from the River Tagus knew what it was, as they had all heard of the swamping of the Peruvian port of Callao and the deaths of up to 10,000 people after a great earthquake in Lima in 1746.

The Lisbon sea wave reached a height of 18 metres (60 feet) – perhaps considerably higher – flooding the streets, squares and gardens up to 180 metres (200 yards) from the waterfront, and returning twice thereafter. The waves swept away the splendid new quay on the river in front of the customs house, along with hundreds of desperate people who had been waiting on the quayside for boats. At the mouth of the Tagus, it hurled boulders weighing as much as 25 tons some 27 metres (90 feet) inland. On the south coast of Portugal, the Algarve, the seabed was exposed in places to a depth of 37 metres (120 feet), and the devastation, except at Faro (which was protected by a lagoon), was so extensive that it was still unrepaired in the early 20th century. In North Africa, the tsunami was the highest ever recorded on the African coast, at 16 metres (52.5 feet), with severe damage and many casualties in Algiers and Tangier.

A British surgeon in Lisbon, Richard Wolfall, who attended the wounded from noon until night on that first day, did his best to evoke the doom-laden atmosphere of the catastrophe, in a letter written on 18 November, which reached London a month later:

> The shocking sight of the dead bodies, together with the shrieks and cries of those, who were half-buried in the ruins, are only known to those who were

> eye-witnesses. It far exceeds all description, for the fear and consternation were so great, that the most resolute person durst not stay a moment to remove a few stones off the friend he loved most, though many might have been saved by so doing: but nothing was thought of but self-preservation; getting into open spaces, and into the middle of streets, was the most probable security. Such as were in the upper stories of houses, were in general more fortunate than those, that attempted to escape by the doors; for they were buried under the ruins with the greatest part of the foot-passengers: such as were in equipages escaped best, though their cattle and drivers suffered severely; but those lost in houses and the streets were very unequal in number to those, that were buried in the ruins of churches . . . all the churches in the city were vastly crouded, and the number of churches here exceeds that of both London and Westminster.
>
> Had the misery ended there, it might in some degree admitted of redress; for though lives could not be restored, yet the immense riches, that were in the ruins, might in some part have been digged out: but the hopes of this are almost gone, for in about two hours after the shock, fires broke out in three different parts of the city, occasioned from the goods and the kitchen-fires being all jumbled together . . . Indeed every element seemed to conspire to our destruction; for soon after the shock, which was near high water the tide rose forty feet higher in an instant than was ever known, and as suddenly subsided. Had it not so done, the whole city must have been laid under water. As soon as we had time for recollection, nothing but death was present in our imaginations.[11]

Even the king, José I, was in dire straits. At the time of the earthquake, he and his family were staying at Belém, on the outskirts of Lisbon, not in their palace in the city, and thereby escaped almost certain death as the palace collapsed. Afterwards, they camped in the garden of the ruined palace. When greeting the British ambassador in Lisbon, the king reportedly told him: 'I was four days ago the richest man here, and now you see me poorer than the meanest beggar in all my dominions – I am without a house, in a tent, without servants, without subjects, without bread.'[12] At the end of November, he decided to build a new lodging out of timber at Belém, because, according to the queen, he had no wish to live ever again in a large stone building with a high ceiling. (In the early 19th century, long after José I's death, the wooden lodging was replaced by a great stone palace.)

The total number of fatalities from the earthquake, fire and tsunami will never be known. In 2004, a mass grave was found under Lisbon's Academy of Sciences Museum, on the site of a former Franciscan monastery. It contained at least 3,000 victims of the earthquake, men, women and children, who had been unceremoniously buried. Some of their bones had fused with sand particles, indicating a fire with a temperature in excess of 1,000 degrees Centigrade (1,800 degrees Fahrenheit); knife marks on a thigh bone may have been the result of cannibalism. Almost certainly, there are other mass graves beneath Lisbon, yet to be discovered. The best estimate of the total number of deaths is 30,000–40,000 in Lisbon, and a further 10,000 in the rest of Portugal, Morocco and Spain, according to Paice. (Recall that 'more than thirty thousand' deaths in Lisbon were mentioned by the *Illustrated London News*, nearly a century after the earthquake.)

All the hospitals in the city had been shaken or burnt to the ground, along with the prisons and the record offices of the municipal authorities, and the king's library containing 70,000 books. Three-quarters of the city's principal religious institutions had gone or were severely damaged; there was no parish church remaining in at least thirty of the forty parishes. Only 3,000 of the city's 20,000 houses were inhabitable. As for the financial cost of the damage, it was of the order of twenty times the value of the cargoes in that autumn's fleets from Brazil, or three times the losses sustained by London in the Great Fire in 1666.

The exact epicentre of the earthquake is also uncertain, despite numerous studies. Undoubtedly, it lay under the Atlantic Ocean – hence the production of the tsunami. One possibility is that the epicentre lay some 200 kilometres (125 miles) west-southwest of Cape Saint Vincent, judging partly by a magnitude-7.3 earthquake in this area in 1969 with a similar (though less intense) spatial pattern of seismicity, which also generated a tsunami. This would locate the epicentre in the Azores–Gibraltar fault zone, near a fault line running from the Azores through the Strait of Gibraltar into the Mediterranean – at the junction of the African and Eurasian tectonic plates. That said, at least four candidates for the epicentre have been proposed, and there is vigorous disagreement between their advocates. Even today, 'the Iberian Peninsula scarcely stands out among the great earthquake zones of the world, or as a likely candidate to produce a great earthquake', seismologists Susan Hough and Roger Bilham observe.[13]

On the Atlantic island of Madeira, a Portuguese colony further to the southwest of Cape Saint Vincent, on the other side of the probable epicentre, the inhabitants heard a rumbling noise in the air, 'like that of empty carriages passing hastily over a stone pavement', and then their houses shook at 9.38 a.m. for about a minute (according to a resident shipper of Madeira wine).[14] But there were no casualties and hardly any damage. At the Rock of Gibraltar itself, on the south coast of Spain, some of the guns of the British batteries were seen to rise, others to fall, as a result of the seismic undulations.

Elsewhere in Europe and North Africa, the shaking was sensible at a distance of 2,400 kilometres (1,500 miles). Its effects were felt over an astonishing area of almost 16 million square kilometres (6.2 million square miles), that is, twice the size of Australia. The earthquake produced seiches (fluctuations in water levels) in the lakes of Britain, including an extraordinary agitation of Loch Ness in Scotland with a substantial wave that threatened a brewery near the loch's waterline. In Finland, 3,500 kilometres (2,200 miles) from the probable epicentre, it disturbed the water at the port of Turku (Åbo).

The tsunami proved to be even more far-reaching. It brought death and destruction to the North African coast, and pandemonium to parts of the coast of Cornwall, the southwestern peninsula of England. On the far side of the Atlantic, in the islands of the Caribbean, the sea retreated as much as 1,500 metres (1,600 yards), beaching a ship that had been floating in 4.5 metres (15 feet) of water in the Dutch Antilles, and then it rose 6.5 metres (21 feet), flooding low land and the upper rooms of houses in the French West Indies.

Of course, those who observed these strange effects had no idea of their cause for a long time, in an age before telegraphs and seismographs.

The news of what had struck Lisbon reached Madrid about a week later, Paris about three weeks later and London only three weeks and three days after the disaster.

Unlike the personal narratives of the English earthquakes published by the Royal Society in London in 1750, data was collected systematically in Portugal – for the first time in the history of earthquakes. An official questionnaire was distributed to parishes by the government. Its thirteen questions covered matters such as the timing and direction of the earthquake; aftershocks and previous earthquakes; the earthquake's effects on bodies of water, including fountains and wells; the size of any fissures; the movements of the sea before the tsunami; the number of deaths; the duration of fires; the damage to buildings; food shortages; and the immediate measures taken by those in authority, whether civil, military or ecclesiastical.

For instance: Did you perceive the shock to be greater from one direction than another? Did buildings seem to fall more to one side than the other? Did the sea first rise or fall? How many hands did it rise above the normal? The answers were stored in the national historical archives in Lisbon, where they can still be read. As historian Charles Davison remarks, the Lisbon quake was the first earthquake 'to be investigated on modern scientific lines'.[15]

The driving force behind this enquiry was the prime minister, who is best known under his later title as the first marquis of Pombal. A dedicated reformer, Pombal was nevertheless paradoxical in his attitude to the disaster, combining enlightenment with authoritarianism. Though sympathetic to science in the investigation of the earthquake, to the secularization of university education and to the application of city planning in the reconstruction of Lisbon, Pombal distrusted the disinterested curiosity of scientists. Though determined to use the disaster to destroy the influence of the Jesuits, Pombal replaced the authority of the church with a virtual dictatorship of his own. And though effective in rooting out corruption in the government, Pombal personally accumulated large holdings of land in Lisbon during the reallocation of property after the earthquake. '[Pombal] wanted to civilise the nation and at the same time to enslave it. He wanted to spread the light of philosophical sciences and at the same time elevate the royal power of despotism', wrote one of his closest collaborators in the area of ecclesiastical and educational reform, António Ribeiro dos Santos, after Pombal's death in 1782.[16] Subsequent generations of Portuguese were divided about Pombal's merits and his reforms. A century and a half passed before he was given national recognition in the form of the great statue that now dominates Lisbon.

Pombal was born in Lisbon in 1699 as plain Sebastião José de Carvalho e Melo, into a family of minor gentry. His father had served in the navy and army, rising to the prestigious post of officer of the court cavalry; his uncle was a landowner, a professor at the University of Coimbra and later arch-priest of the Lisbon patriarchy. Only in 1769, late in life, did Pombal acquire his title. So his ancestry was 'neither as grand as his title might imply, nor as modest as his enemies claimed', writes his biographer, Kenneth Maxwell.[17]

From 1739 to 1743, he served as the Portuguese ambassador in London. He was offended by the way in which Britain casually took for granted its dominance over Portugal. Nevertheless, by moving in the circle of the Royal Society, Pombal set out to explore the origins and maintenance of British commercial and naval power in detail, and formed an appreciation that helped to inspire his later Portuguese reforms. But he would always put political considerations before any others. Thus, after the 1755 earthquake, Pombal regarded the Jesuits purely as a reactionary political force in Portugal, and overlooked their interest in educational reform and new ideas, including science; and later, in the 1760s–70s, he suspiciously assumed that the British voyages of Captain James Cook were primarily directed against the Portuguese and Spanish dominions in South America, rather than being motivated by geographical and scientific curiosity about the Pacific and the unknown southern continent.

Back in Portugal, Pombal became the power behind the throne in 1750, when José I was crowned king and appointed Pombal as his secretary of state. In 1755, he assumed the role, *de facto*, of prime minister. Thereafter, Pombal enjoyed the king's unqualified support for more than two decades, until José's death in 1777.

The earthquake was the making of Pombal. In its bewildering and frightening aftermath, when the impractical king asked his practical prime minister what he should do now, Pombal is famously supposed to have given the blunt advice: 'Bury the dead and feed the living.'[18] As noted by Pombal's biographer: 'He took quick, effective, and ruthless action to stabilise the situation.' Looters were hanged. To prevent disease, 'bodies of the earthquake victims were quickly gathered and with the permission of the Lisbon patriarch, taken out to sea, weighted and thrown into the ocean' – or buried beneath Lisbon in unmarked mass graves, as became clear only in 2004. 'Rents, food prices, and the cost of building materials were fixed at pre-earthquake levels. No temporary rebuilding was permitted until the land was cleared and plans for new construction drawn up.'[19]

Less immediately, Pombal had to deal with clerical arguments that the devastation was a divine response to the city's moral failings. The clergy had been put in an especially invidious position by the earthquake. If it truly was divine punishment for the sins of Lisboetas, then why had it occurred on All Saints' Day, destroyed so many religious institutions and killed so many clergy? Although official accounts mention only a few hundred deaths of clerics, friars and nuns – some of whose bodies were supposedly found under the rubble in a state of miraculous preservation – the true figure was unquestionably far greater, given the clergy's preponderance in Lisbon at the time; the real number of clerical fatalities must have been suppressed by the church authorities. Theologically, the earthquake presented a profound paradox.

Voltaire was incensed by such religiosity. A few days after hearing the first news of the Lisbon disaster, he informed a banker friend in Lyons on 24 November:

> What a sad game of chance the game of human life is! What will the preachers say, especially if the palace of the Inquisition remains standing? I flatter myself at least that the reverend Fathers, the Inquisitors, will have been crushed like all the others. That ought to teach men not to persecute men, for while some holy scoundrels burn a few fanatics the earth swallows up the lot of them whole.[20]

But Voltaire also disagreed with the optimistic outlook of secular society. Optimism as a philosophy sprang from the influential ideas of philosopher and mathematician Gottfried Leibniz and poet and essayist Alexander

Pope. Leibniz had conceived the world to be 'the best of all possible worlds' in a celebrated essay on good and evil published in 1710. Pope, in a well-known poem of 1734–5, *An Essay on Man*, had declared:

All Nature is but Art, unknown to thee;
All Chance, Direction, which thou canst not see;
All Discord, Harmony not understood;
All partial Evil, universal Good.
And, spite of Pride, in erring Reason's spite,
One truth is clear, 'Whatever is, is RIGHT'.[21]

In Voltaire's poem about the earthquake, published anonymously in Paris in January 1756, he therefore questioned how either the church authorities or optimist philosophers could possibly justify the destruction of Lisbon. Why not decadent London or Paris? Why does Lisbon lie in ruins, while in Paris they dance? he asked. In 1759, in the Lisbon earthquake section of *Candide*, Voltaire satirized three typical reactions to the tragedy: that of the common man, the optimist philosopher (Doctor Pangloss) and the innocent (Candide):

Scarcely had they set foot in the city . . . than they felt the earth quake beneath their feet. In the port a boiling sea rose up and smashed the ships lying at anchor. Whirlwinds of flame and ash covered the streets and public squares: houses disintegrated, roofs were upended upon foundations, and foundations crumbled.

Thirty thousand inhabitants of both sexes and all ages were crushed beneath the ruins. The sailor said with a whistle and an oath: 'There'll be some rich pickings here.'

What can be the sufficient reason for this phenomenon?' wondered Pangloss.

'The end of the world is come!' Candide shouted.[22]

Jean-Jacques Rousseau, another philosopher, disagreed with Voltaire's argument against optimism on the reasonable grounds that people in Lisbon had brought disaster upon themselves by choosing to invest in fragile urban buildings and packing them with valuable possessions, which they hesitated to abandon, rather than living in safe small houses in natural surroundings, from which they could easily escape. (Presumably, José I would have agreed.) And some other thinkers, such as Immanuel Kant, argued that great earthquakes might have hidden benefits. But for the majority of thinking Europeans, Voltaire's criticism marked the beginning of an irreversible shift away from both religious explanations of natural disasters and the philosophy of optimism.

The poet Johann Wolfgang von Goethe was one of them. As an adult, Goethe recalled that the horror of the earthquake had startled a more or less quiet and happy society and profoundly disturbed his own peace of mind, as a child of six in 1755:

> God, the creator and preserver of heaven and earth, God, said to be omniscient and merciful, had shown himself to be a very poor sort of father, for he had struck down equally the just and the unjust. In vain [my] young mind sought to combat this idea; but it was clear that even learned theologians could not agree about the way in which to account for such a disaster.[23]

Portugal's prime minister, Pombal, was another. He had no truck with the Jesuitical account of the earthquake. Soon after it occurred, the Portuguese government issued decrees to forbid priests from stirring up feelings of recrimination and guilt in the population. When the leader of the Portuguese Jesuits, Italian-born Gabriel Malagrida, prophesied a second great earthquake in November 1756, Pombal banished him from the court. After a power struggle in which there was a mysterious attempt on the life of the king, Pombal had the Jesuit confessors to the royal family expelled from their positions and replaced by priests in his own confidence. In September 1759, the entire Jesuit order was expelled from Portugal and its empire, and any communication between Jesuits and Portuguese subjects – whether verbal or in writing – was prohibited. In 1760, Portugal and the Vatican broke off diplomatic relations. The year after, Malagrida was arrested, tried for high treason and conveniently convicted of heresy by the Inquisition

– now led by Pombal's brother. After a public garrotting, Malagrida's corpse was burnt at the stake and his ashes thrown into the Tagus. Small wonder that in 1982, the bicentenary of Pombal's death, a Jesuit magazine asserted that: 'His methods were in effect an anticipation and mixture of the methods of Goebbels and Stalin.'[24]

Meanwhile, Pombal saw the destruction of Lisbon as a practical challenge. From the late 1750s, the increasingly dictatorial Pombal pursued the rebuilding of Lisbon, partly funded by imposing an extra 4 per cent import tax (over the strong objections of British merchants in Lisbon). Indeed, 'he seized on the disaster as an opportunity for urban development and commercial expansion', notes historian Kevin Rozario in *The Culture of Calamity*:

> Employing the absolute power of the monarchy and the dwindling but still fabulous riches of empire, Pombal built a magnificent new metropolis, and he overhauled the economic and political structure of his country. Outlawing unauthorized construction in central Lisbon, Pombal appointed military engineers to develop a coherent blueprint for the city, and within months hills were flattened, a confusion of narrow alleyways had given way to a grid of wide thoroughfares, and a system of sewers had appeared beneath the streets. New buildings were safer and more uniform in appearance, constructed out of standardized building materials and built around flexible, earthquake-proof frames designed to move with the shaking of the earth.[25]

The rebuilding process continued well beyond Pombal's own fall from power in 1777 after the death of his royal patron, José I, and the occurrence of two further earthquakes in 1796 and 1801, into the 19th century. But now Lisbon suffered again, this time from both French and British occupations during the Napoleonic wars, in which the Portuguese royal family was forced to flee to Brazil in 1807.

Although, over the decades, the city recovered much of its former prosperity, the loss of Brazil as a colony in 1822 was a serious blow from which Portugal would never recover. From this time onwards, Lisbon was haunted by a sense of deep loss – expressed in the sorrowful Portuguese singing known as *fado* ('fate'), which began to be heard on Lisbon's streets from the 1820s. As Dickens noticed on his visit in 1858, for all the colour, charm and exuberance of the city's street life, the 'common people' of Lisbon never seemed to laugh.[26]

CHAPTER

4 BIRTH OF NATIONS: CARACAS, 1812

Republican Simón Bolívar (in shirt sleeves) confronts royalist José Domingo Diaz in the earthquake-ruined cathedral of Caracas in 1812.

Brazil's declaration of independence from Portugal in 1822 after three centuries of colonial rule was part of a continent-wide movement towards political independence in Latin America. In 1800, just before the beginning of the Napoleonic wars, virtually all of Central and South America – from northern Mexico to southern Argentina and Chile – was under Spanish or Portuguese royal rule. By 1825, this dominion had been swept away, except in Cuba and Puerto Rico, leading to the creation of new nations: Bolivia, Colombia, Ecuador, Peru and Venezuela.

Although the first of the political earthquakes in this storm struck Haiti, when it gained its independence from France in 1804, and was followed by others in Argentina in 1816 and Mexico in 1821, the most politically influential one occurred in the young republic of Venezuela, whose independence from Spain had been declared in July 1811 under the leadership of General Francisco de Miranda and Simón Bolívar. Just before Easter, on 26 March 1812, Venezuela's capital, Caracas, was severely damaged by a physical earthquake that would generate political aftershocks for the next two decades. Indeed, a modern biographer of Bolívar, John Lynch, chooses to begin his biography with the Caracas earthquake of 1812.

When the quake struck, a royalist chronicler, José Domingo Diaz – no supporter of the fledgling republic – was present. Diaz recorded his first strong impressions as follows (in Lynch's translation from the Spanish):

> It was four o'clock, the sky of Caracas was clear and bright, and an immense calm seemed to intensify the pressure of the unbearable heat; a few drops of rain were falling though there was not a cloud in the sky. I left my house for the Cathedral and, about 100 paces from the plaza of San Jacinto and the Dominican priory, the earth began to shake with a huge roar. As I ran into the square some balconies from the Post Office fell at my feet, and I distanced myself from the falling buildings. I saw the church of San Jacinto collapse on its own foundations, and amidst dust and death I witnessed the destruction of a city which had been the admiration of natives and foreigners alike. The strange roar was followed by the silence of the grave. As I stood in the plaza, alone in the midst of the ruins, I heard the cries of those dying inside the church; I climbed over the ruins and entered, and I immediately saw about 40 persons dead or dying under

> the rubble. I climbed out again and I shall never forget that moment. On the top of the ruins I found Don Simón Bolívar in his shirt sleeves clambering over the debris to see the same sight that I had seen. On his face was written the utmost horror or the utmost despair. He saw me and spoke these impious and extravagant words: 'We will fight nature itself if it opposes us, and force it to obey.' By now the square was full of people screaming.[1]

Bolívar's own recollection of this meeting was decidedly different from that of Diaz. The earthquake woke him from a siesta. Then:

> I immediately set about trying to save the victims, kneeling and working towards those places where groans and cries of help were coming from. I was engaged upon this task when I saw the pro-Spanish José Domingo Diaz, who looked at me and commented with his usual scorn: 'How goes it, Bolívar? It seems that nature has put itself on the side of the Spaniards.' 'If nature is against us, we will fight it and make it obey us,' I replied furiously.[2]

In truth, most of the events in this catastrophe are disputed. The magnitude of the earthquake that struck Caracas has been tentatively estimated by two seismologists as between 6.9 and 7.2; but seismologists disagree whether there was a single massive earthquake or up to four smaller earthquakes at different places in Venezuela. Historians disagree about the effects, because of the earthquake's timing during a war at a turning point between colonial rule and independence. It is 'complicated by heroic and nationalistic reconstructions of history', according to a Venezuelan anthropologist, Rogelio Altez, who recently compared the various accounts of the earthquake published at the time and later in the 19th century.[3]

For example, one report dated 30 March 1812, by a British naval captain stationed on the nearby Caribbean island of Curaçao, estimated 15,000–20,000 deaths in Caracas; another dated 9 April, from Caracas itself, mentioned only 1,000 deaths; while the archbishop of Caracas gave a figure of 10,000–12,000 deaths, based on reports from his parish priests, although this figure appears to have included deaths in the areas surrounding Caracas. The total population of Caracas at the time is also unclear, ranging from 31,813 (according to Diaz) to 50,000 people. As for contemporaneous estimates of the damage, these range from nine-tenths of the city destroyed to only one-third of the buildings having fallen (the estimate of the archbishop). The scientist-explorer Alexander von Humboldt, in his *Personal Narrative* of his travels in tropical Spanish America (an inspiration for Charles Darwin in his voyage on HMS *Beagle*), claimed that 'Caracas was entirely overthrown' by the earthquake, although he was not personally present and relied on the reports of others, notably Louis Delpeche, a French official stationed in the city, who published an article on the earthquake in a Paris-based journal in 1813, estimating the earthquake fatalities as 9,000–10,000 people.[4]

Altez, after analysing the deaths recorded in two 1812 funeral books from parishes in the urban area of Caracas and another funeral book from a parish on the outskirts of the city, came up with what is probably the most reliable estimate for the nineteen parishes within the urban perimeter: close to 2,000 fatalities. He was, however, unable to estimate the true level of destruction accurately. But Altez does conclude that – despite Diaz's dramatic description of the deaths in the ruined church – more deaths resulted from the collapse of houses than from religious buildings, because the tile roofs of the houses were high and very heavy, and propped up by weak, thin wooden supports. In other words, there were more earthquake victims among the maids, children and slaves left behind at home than among the middle- and upper-class families attending divine service in the cathedral on Maundy Thursday (in contrast to the slaughter of worshippers under disintegrating churches in Lisbon in 1755). Bolívar's own house was severely damaged; its floors, doors and windows were wrecked. There was also rapid and extensive looting of the ruins, in which the poor robbed gold from corpses and ripped jewelry from the ears of trapped women crying out for help. Smoke from fires and thick yellow dust from collapsed buildings, combined with the darkness of night, concealed the activities of the looters.

Naturally, not only Caracas was impacted by the shaking. Indeed, the quake (or quakes) affected the entire upper half of South America. In Venezuela, there was heavy damage from Caracas to Mérida, over a distance of about 500 kilometres (320 miles). It damaged other cities also controlled by the republicans: Barquisimeto, La Guaira, San Carlos, San Felipe and

Trujillo. In the port of La Guaira, only one building was still standing: that of Spain's formerly all-powerful Royal Guipuzcoana Company. In Barquisimeto, an entire regiment of 1,500 men supposedly fell into a fissure and perished. In San Felipe, the 600-strong garrison was annihilated by the collapse of its barracks. Estimates of total fatalities in all cities, towns and villages range from 20,000 (the figure favoured by Humboldt) to 120,000 people, although these figures are inevitably not dependable, given the uncertainty about the deaths in Caracas, where records were fuller than in the rest of the country, despite their considerable destruction. Even Cartagena, the chief city of faraway New Granada (modern-day Colombia), reported crippling damage.

By chance, the damage was worst in the areas controlled by Miranda, Bolívar and the rebels, who called themselves 'patriots'. Royalist strongholds, such as Coro, Guayana, Maracaibo and Valencia, were untouched. The local Catholic authorities – including royalists like Diaz – had a field-day with this coincidence, combined with the occurrence of the earthquake just before Easter. The archbishop of Caracas, Narciso Coll y Prat, thundered that the earthquake was terrifying but well-deserved punishment for Venezuelan vices, notably the city's 'patriot' disloyalty to Spain, and invoked the biblical warnings about Sodom and Gomorrah. Bolívar, while organizing republicans to rescue the dying and the dead by digging with their bare hands and removing the victims on makeshift stretchers, stumbled across a red-faced priest berating a frightened crowd. 'On your knees, sinners!' the priest shouted. 'Now is your hour to atone. The arm of divine justice has descended on you for your insult to his Highest Majesty, that most virtuous of monarchs, King Ferdinand VII!'[5] Legend has it that Bolívar threatened the priest with his sword.

Without doubt, this religious exploitation of the earthquake was effective in a highly superstitious land at stirring up a counter-rebellion against the republic. 'The Venezuelan people, terrified by these revelations and sure now that God had spoken, streamed to the royalist side', writes another Bolívar biographer, Marie Arana. 'Republicans deserted to the king's army. As Spain's General Monteverde advanced swiftly towards the republic's capital, he had no trouble recruiting troops.'[6]

In July 1812, soldiers under Bolívar's command were compelled by the treasonable action of one of their officers to evacuate a key strategic position, the coastal fortress at Puerto Cabello, after Miranda proved unable to send the embattled Bolívar reinforcements. On 25 July, Miranda, without consulting Bolívar, signed surrender terms with Monteverde, not only because of his weak military position but also because of an uprising of blacks to the east of Caracas, the effects of the earthquake and a shortage of provisions in Caracas. Miranda prepared to escape Venezuela for Curaçao from the port of La Guaira on board a waiting British ship. However, he had failed to take account of Bolívar and some other republican leaders, who condemned his actions as highly treasonous. At the last minute, while Miranda was asleep in the Guipuzcoana building at La Guaira, this group came at night to arrest him. Grasping his aide's lantern, Miranda held it high so that he could study the face of Bolívar and others. 'Ruffians! Ruffians!' he sighed. 'All you know is how to make trouble.'[7] He was chained and soon handed over to the Spanish, who imprisoned him until his miserable death in a dungeon in Cadiz in 1816. Bolívar had apparently wanted to shoot Miranda, but was restrained by his fellow leaders. In the confusion that followed the arrest, Bolívar managed to escape in disguise and ride through the night to Caracas.

For this morally dubious act, Bolívar was rewarded by the Spanish, after the impassioned intercession of an old family friend, a royal official who was on excellent terms with General Monteverde. 'Alright,' the general told his secretary, while eyeing the unwelcome Bolívar, 'Issue this man a pass as a reward for services rendered the King when he imprisoned Miranda.' Unable to restrain himself, Bolívar replied: 'I arrested Miranda because he was a traitor to his country, not in order to serve the King!'[8] A piqued Monteverde threatened to withdraw the travel pass, but after further intercession by the go-between, it was granted. In late August, Bolívar left Venezuela on a Spanish ship headed for Curaçao. There his baggage was confiscated by an unfriendly British governor and he became a refugee. By late October, he had settled in Cartagena, a port keen on free trade, which had declared its independence from Spain in late 1811.

Even now, notes historian Robert Harvey in *Liberators*, his study of the independence movement in South America, Latin American historians, who venerate both Bolívar and Miranda, 'find it difficult to come to terms with the fact their greatest hero betrayed their second-greatest into captivity. The two are still portrayed in popular iconography as the closest of friends.'[9] In Harvey's view, Bolívar should be more accurately regarded as 'the sorcerer's apprentice'.[10]

The First Republic of Venezuela was finished, but Bolívar – unlike Miranda – would live to fight again. From its ruins, 'emerged the unmistakable signs of a leader: a commander's ruthlessness, an inner fortitude, a resolution in the face of adversity, and an ability to pick himself up from calamity and come back fighting', writes Lynch. 'The Spanish American revolution re-enacted the scene many times in the next 20 years: his individual survival amidst collective failure.'[11]

In Cartagena, Bolívar wrote his first major political text, the trenchant 'Cartagena Manifesto', dated December 1812. Besides defining the course of his future career as an authoritarian liberator, it set out, precisely, his own analysis of the historical importance of the Venezuelan earthquake nine months earlier. It begins:

> I am, Granadans, a son of unhappy Caracas who miraculously escaped from amid her physical and political ruins . . .
>
> Allow me, inspired by a patriotic zeal that emboldens me to address you, to sketch for you the causes that led Venezuela to her destruction . . .
>
> The most grievous error committed by Venezuela as she entered the political arena was undoubtedly her fatal adoption of the governing ideal of tolerance, an ideal immediately rejected as weak and ineffective by everyone of good sense, yet tenaciously maintained right up to the end with unparalleled blindness.[12]

When Bolívar comes to the earthquake – evidence of which lay around him in Cartagena – he argues that:

> It is true that the earthquake of 26 March was as devastating physically as it was spiritually and can fairly be said to have been the immediate cause of Venezuela's ruin, but this catastrophe would not have produced such fatal effects if Caracas had been governed by a single authority that could have quickly and vigorously set about repairing the destruction, without the complications and conflicts

> that slowed down the recovery in the provinces, exacerbating the harm until it was incurable.
>
> If instead of a languid and untenable confederation Caracas had established a simple government as required by her political and military situation, you would still exist today, Venezuela, and would be enjoying your freedom!
>
> Following the earthquake, the ecclesiastical influence was a considerable factor in the insurrection of the smaller towns and cities and in the introduction of the enemy into the country, abusing the sanctity of its ministry most sacrilegiously in favour of those fomenting civil war. Still, in our naivety we have to confess that these treacherous priests were encouraged to commit the execrable crimes they have been accused of because impunity for crimes was absolute . . .
>
> From the foregoing, it is clear that among the causes leading to the fall of Venezuela, first was the nature of her constitution, which was, I repeat, as inimical to her interests as it was favourable to those of her enemies. Second was the spirit of misanthropy that took hold of our governors. Third was the opposition to the establishment of a standing army that could have served the Republic and warded off the blows dealt by the Spaniards. Fourth was the earthquake accompanied by the fanaticism that gave such dire interpretations to this event. Finally, there were the internal factions that were in reality the mortal poison that pushed the country into her grave.[13]

In Bolívar's view, Venezuela required a degree of dictatorship, not only to recover from the earthquake but also to survive as a nation. (No doubt, the marquis of Pombal felt the same way, regarding the recovery of Portugal in the 1750s.) It was a view he would come to hold with increasing conviction during his leadership of the liberation movement from Spanish rule for other new nations in South America. Not as much dictatorship as he had observed in the France of the Emperor Napoleon I – but certainly more than he had seen in the United States of America.

In Paris, in 1804, Bolívar had witnessed the coronation of Napoleon Bonaparte with deep dismay. As he would later recall to one of his biographers:

> I regarded the crown that Napoleon placed on his head as a miserable, outdated relic. For me, his greatness was in his universal acclaim, in the interest

> his person could inspire. I confess that the whole thing only served to remind me of my own country's enslavement, of the glory that would accrue to him who would liberate it. But I was far from imagining that I would be that man.[14]

In the US, for several months in 1807, Bolívar had lived in Charleston, in the slave-owning south, and in Philadelphia, in the north, where the US constitution had been adopted in 1787, and perhaps in other cities. He concluded that the North American political system was not suitable for South America. For Venezuelans suffered from complex, fissiparous divisions of race and class and lack of political experience, all of which had been fostered by centuries of Spanish colonial rule, such that the Creole aristocracy by and large preferred even a foreign yoke to domination by Americans of an inferior class. Bolívar, though himself born into the Creole aristocracy, did not share this particular attitude. But he recognized that Venezuelan independence based on democracy would descend into chaos in the hands of this group when in power. They were not ready for the federal constitution actually adopted by the republic in 1811, which mimicked that of the US government. 'The federal system, although it is the most perfect and the most suitable for guaranteeing human happiness in society, is, notwithstanding, the form most inimical to the interests of our emerging states', Bolívar asserted at Cartagena.[15]

As for a Venezuelan standing army, this, too, was a necessity because Venezuelans, unlike the citizens of earlier nascent republics who had thrown off the yoke of tyrants, beginning with those of ancient Greece, lacked 'political virtues, Spartan habits and military character'.[16] North America's victory in its struggle for independence from Britain in the 1770s–80s without a pre-existing standing army or any contingent of veterans was a problematic example, Bolívar conceded, which he put down to North America's long-standing protection by the Atlantic and Pacific Oceans against outside invasion.

By strange coincidence, North America was struck by major earthquakes at almost the same moment as South America. The differing responses of the United States and Venezuela are revealing.

In the sparsely populated mid-continental area of Missouri, centred around New Madrid, three major North American earthquakes on

16 December 1811, 23 January 1812 and 7 February 1812 created waterfalls in the Mississippi River and, astonishingly, caused the flow of the river to reverse. One eyewitness wrote that: 'A bursting of the earth just below the village of New Madrid arrested this mighty stream in its course, and caused a reflux of its waves, by which in a little time a great number of boats were swept out by the ascending current into the bayou, carried out and left upon the dry earth.'[17] Another observer, living in Louisville, Kentucky, catalogued more than 600 separate felt earthquakes between 26 December 1811 and 23 January 1812 according to his own estimate of their intensity. The persistent legend that the Missouri quakes even caused the church bells of far-off Boston in Massachusetts to peal is untrue, as proved by the lack of any mention of the quake in contemporary newspapers in the Boston area; but they did set church bells ringing in Charleston, South Carolina, nearly 1,000 kilometres (650 miles) away from New Madrid. Their magnitude was not as great as 8.0–8.75, as formerly believed by seismologists; they were probably in the range 7.4–8.1, or even as low as 7.0. Yet, the New Madrid sequence 'remains the most dramatic example of an extended earthquake sequence with multiple large mainshocks ever witnessed in the United States', notes Susan Hough.[18]

The North and South American earthquakes, and other geological events in the period 1811–12 – such as the spectacular eruption of a volcano (La Soufrière) in the Caribbean, and the sudden appearance and disappearance of a volcanic island (Sabrina) in the Azores – may have been more than a coincidence. After all, suggested Humboldt, if the tremors from the 1755 Lisbon earthquake could be felt almost simultaneously on the coasts of Finland, at Lake Ontario in Canada and on the island of Martinique in the Caribbean, was it unreasonable to imagine that Missouri and Caracas might be 'simultaneously agitated by commotions proceeding from the same centre of action'?[19] Two centuries later, there is no proof that Humboldt's conjecture was correct, but neither can the possibility of a geological interconnection be conclusively excluded. The cause of the New Madrid intraplate earthquakes – far away from North America's plate-tectonic boundaries on the Pacific coast and beneath the Atlantic Ocean – is still puzzling to seismologists, despite extensive scientific study, as we shall see.

Non-scientists reacted more predictably to the US earthquakes. Before the first one, a large comet blazed in the skies of North America from September 1811 for some four months; it was visible during the first earthquake in December 1811, but had disappeared by the time of the second and third quakes. The influential Shawnee chief, Tecumseh, was then travelling the Midwest in a crusade designed to stir up Indian tribes against the US government's confiscation of their lands. In a speech to the Osage tribe, Tecumseh seems to have taken the comet and the earthquakes as portents: 'Brothers, the Great Spirit is angry with our enemies. He speaks in thunder, and the earth swallows up villages, and drinks up the Mississippi. The great waters cover the lowlands. Their corn cannot grow, and the Great Spirit will sweep those who escape to the hills from the earth with his terrible breath.'[20] This statement was presumably made in late 1811 or early 1812, but is sometimes dated to early 1811, before the earthquakes, and is therefore known as 'Tecumseh's prophecy'. Yet if this earlier date is actually correct, 'Tecumseh would be perhaps the only individual since the dawn of time to have made a successful earthquake prediction . . . *and not bothered to take credit for it*', in the sceptical words of seismologists Hough and Bilham.[21]

Among white Americans, a Connecticut newspaper quoted by the *Pittsburgh Gazette* in April 1812, was almost equally portentous: 'We have within a few years seen the most wonderful eclipses, the year past has produced a magnificent comet, the earthquakes within the past . . . months have been almost without number – and . . . we constantly "hear of wars and summons of wars" . . . "Can ye not discern the signs of the times."'[22] Preachers and churches did well out of the earthquakes, as they would in Caracas after the March earthquake. In the region around New Madrid, despite its small population of white settlers, the Methodist church is said to have acquired more than 15,000 new members. Furthermore, in June 1812, the US embroiled itself in a war with Britain. This would include the British burning of the capital, Washington, and would last until early 1815.

However, unlike the earthquake in Venezuela, there was never the slightest risk to the stability of the United States from the Missouri earthquakes. One reason, fairly obviously, was the midwestern location of the epicentres far from the centres of power on the east coast; indeed, the affected regions became part of the Union only in 1821, and received no financial relief from the US Congress until two years after the earthquakes. But also important was the unity of the US political leadership, unlike the fatally divided leadership in Venezuela.

During 1812, the United States gave $50,000 in aid to the victims of the earthquake in Venezuela. Flour was sent to Caracas in five ships: 'a valuable pledge of the mutual sympathy which ought for ever to unite the nations of North and South America', thought Humboldt.[23] But this practical support did not translate into North American political support for South American freedom. There would be no US military assistance for the colonies' fight against Spain. Back in 1786, Thomas Jefferson had privately noted that Spain's South American colonies were ripe for the plucking, possibly by the US. By the end of the century, though, there was less taste for this idea. President John Adams, with Jefferson as his vice-president, noted that: 'You might as well talk about establishing democracies among the birds, beasts, and fishes as among the South American people.'[24] In 1806, President Jefferson and his secretary of state, James Madison, were embarrassingly implicated in a botched attempt by a US military officer, the son-in-law of Adams, to help Miranda seize power in Venezuela. Neither Jefferson nor his successor, President Madison, had any involvement in the declaration of the First Republic of Venezuela in 1811. Indeed, Madison did not formally recognize the new republic, although he did urge the US Congress to support the republic with arms. Its only response was an unemotional statement to the effect that revolutions and declarations of independence were insufficient evidence of nationhood. By 1815, when Adams was eighty, he had become so disenchanted with South America's urge for independence that he wrote scathingly:

> What could I think of revolutions and constitutions in South America? A people more ignorant, more bigoted, more superstitious, more implicitly credulous in the sanctity of royalty, more blindly devoted to their priests, in more awful terror of the Inquisition, than any people in Europe, even in Spain, Portugal, or the Austrian Netherlands, and infinitely more than in Rome itself.[25]

Privately, Bolívar might well have agreed with Adams's verdict on his own people. In the 1812 'Cartagena Manifesto', he publicly stated that the government of the First Republic of Venezuela 'ended up in the hands of men who were incompetent, corrupt, or uncommitted to the cause of independence. The party spirit prevailed in all matters, causing more chaos than the events themselves. Our division, not the Spanish forces, reduced us to slavery.'[26] By the 1820s, Bolívar came to believe that Latin Americans 'were not ready for a truly democratic government: abject, ignorant, suspicious, they did not understand how to govern themselves, having been systematically deprived of that experience by their Spanish oppressors', according to his biographer, Arana.[27] Only presidents for life and dictatorship seemed to work in South America, Bolívar concluded, with sadly accurate foresight into the 19th- and 20th-century history of the continent. That said, 'he never regarded dictatorship as a long-term solution,' notes a modern editor of Bolívar's writings, David Bushnell, 'invariably assuming that in due course limited, constitutional government (ideally in line with his recommendations) would be restored.'[28]

Intelligent, wide-ranging and emotionally stirring as his political writings were, Bolívar was first and foremost a man of action – like Napoleon Bonaparte, as is obvious from a brief outline of his career, following the events of 1812 in Caracas. While in New Granada, Bolívar was named commander of an expeditionary force intended to liberate Venezuela. After a sweeping campaign, in six pitched battles he defeated the Spanish, re-entered Caracas and founded the Second Republic of Venezuela in August 1813. On his orders, a thousand chained Spanish prisoners were beheaded. But the following year, Bolívar was routed by a cavalry force of royalist irregulars, former cowboys (*llaneros*) led by the brutal José Tomás Boves, who subjected Caracas to even more awful atrocities. Bolívar narrowly escaped the fate of Miranda and was eventually forced to flee into exile in Jamaica. There, in mid-1815, he wrote the most important document in his career, 'The Letter from Jamaica'. This proposed constitutional republics throughout Latin America, on the model of the government of Great Britain, with a hereditary upper house, an elected lower house and a president chosen for life. Meanwhile, Spain sent a strong expeditionary force across the Atlantic.

Bolívar, unable to secure support from Britain or the US, turned for support to the republic of Haiti, which supplied him with money and weapons. In 1817, he established a headquarters in the Orinoco region of Venezuela, with its capital at Angostura, where he was relatively safe from Spanish attack. He engaged the services of several thousand mercenaries, mostly British and Irish. From there, with their help, in 1819 he launched a daring surprise attack on the Spanish forces in New Granada via routes that the Spanish had considered impassable, crossing flooded plains and icy mountains. After the crucial battle of Boyacá, most of the royalist army surrendered, and Bolívar entered Bogotá.

This was a turning point in the wider struggle against Spain in northern South America. In 1821, Venezuela was finally freed from Spanish rule by Bolívar at the battle of Carabobo. The following year, Ecuador was secured at the battle of Pichincha by a lieutenant of Bolívar, Antonio José de Sucre. Two years later, Bolívar scored a major victory in the Peruvian highlands at the battle of Junín, followed by Sucre's defeat of the Peruvian viceroy at the battle of Ayacucho, after which Upper Peru took the name Bolivia. Since Argentina had declared independence in 1816, Chile in 1818 and Brazil in 1822 – under different leaders – by 1825 almost the entire South American continent was politically independent.

From then until Bolívar's death in 1830, there were bitter disputes between him and other leaders, including an assassination attempt and his resignation as president of Gran Colombia: the federation of Bolivia, Colombia, Ecuador, Panama, Peru and Venezuela, plus western Guyana and northwestern Brazil. But there can be no dispute about his historic importance as a liberator. Equally clear is the role of the 1812 earthquake in pushing him down this path of liberation, beyond his native Venezuela. As Bolívar recognized, without the earthquake's ruin of Caracas and other republican cities, and the royalist backlash against the republican cause that followed the quake, the First Republic of Venezuela might have survived, and he would have remained there at its head. In which case, the liberation of the rest of northern South America from Spanish colonial rule – lacking Bolívar's seismic leadership – would probably have taken much longer than it actually did.

CHAPTER

5 SEISMOLOGY BEGINS: NAPLES, 1857

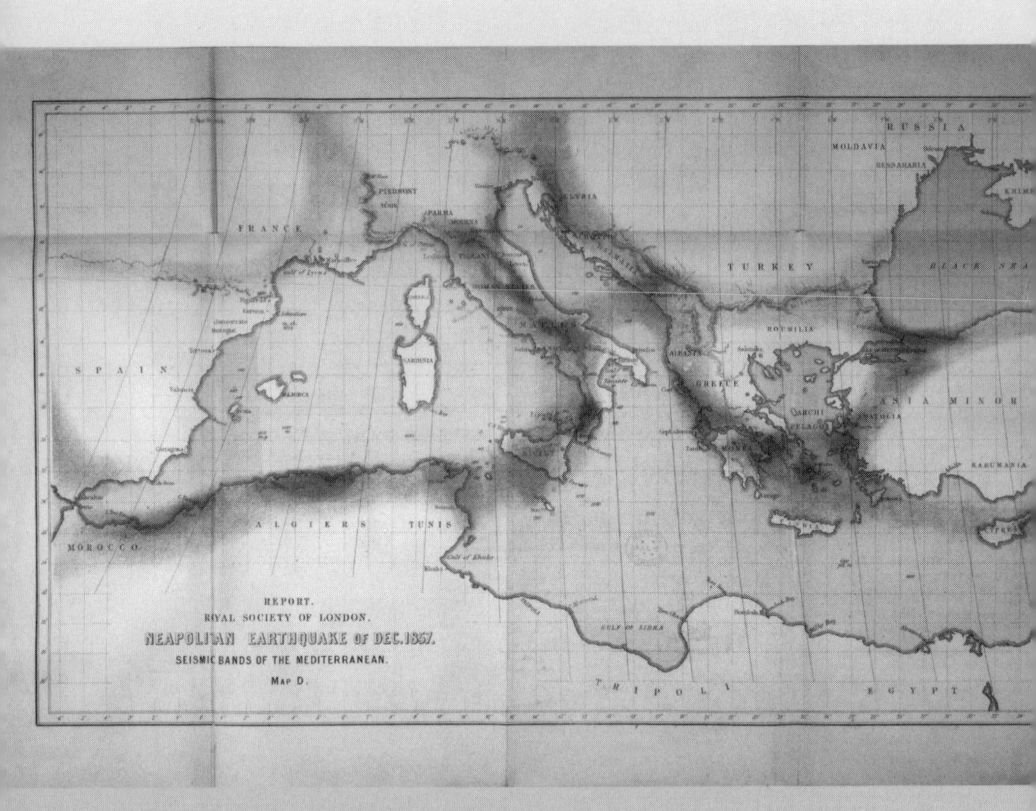

Seismic bands of the Mediterranean by Robert Mallet, published in his report, *Great Neapolitan Earthquake of 1857: The First Principles of Observational Seismology.*

Although supernatural explanations of earthquakes were still prevalent during the early 19th century – as in Caracas in 1812 – science had by then made some progress in understanding seismicity, without as yet becoming the discipline of seismology. The starting point dates, as we know, from the 1750s: the 1750 earthquakes in London, the 1755 earthquake in Lisbon and the geological paper based on their data published by astronomer John Michell in the Royal Society's *Philosophical Transactions* in 1760, in which Michell correctly identified the wave motion involved in earthquakes. The same idea was proposed, if less articulately, by another astronomer, on the other side of the Atlantic, John Winthrop IV of Harvard University, when he saw the bricks in his chimney in Boston move in an undulation: rising up in sequence, then quickly dropping back into place. Winthrop described the motion as 'one small *wave of earth* rolling along'.[1] The strange event happened in 1755 after an undersea earthquake off Cape Ann, north of Massachusetts Bay, just seventeen days after the Lisbon earthquake. (Whether the earthquake in Europe might have triggered the earthquake in North America is unknown, but not out of the question.)

However, Michell's and Winthrop's mid-18th-century insights would have no influence on the further study of earthquakes until the mid-19th century. Theories about seismic waves were all very well, but first it was necessary to try to measure earthquakes: by carefully examining and classifying the destruction they caused, and in due course by monitoring them with seismometers and seismographs of increasing sensitivity and accuracy.

In 1783, the world's first touring earthquake commission was appointed as a result of six disastrous quakes in Calabria, the 'toe' of Italy south of Naples, in February–March of that year. These claimed some 35,000 lives – including six members of the family of the Neapolitan secretary of war – and were massively destructive. But unusually, the destruction was localized: some towns were flattened, while neighbouring towns escaped with only light damage. This variation in levels of destructiveness provided investigators with valuable data that permitted the first attempts to measure and compare earthquakes.

The secretary of war toured the stricken area and noted that the biggest of the six earthquakes, on 28 March, which was also the last in the series, nevertheless was not the most lethal quake. A possible explanation, he thought, was that by this date the inhabitants of the region, terrified by the earlier deaths and destruction, had moved out of doors, away from buildings. Following in his footsteps, the Academy of Sciences and Fine Letters of the kingdom of Naples surveyed more than 150 towns and villages. Its 372-page report, containing maps and drawings, tabulated the time of each earthquake, the number of fatalities and the level of damage, any aftershocks and sea waves, the impact on survivors and any subsequent epidemics, as well as describing the geology of the area.

No theory of earthquakes emerged from this report, but it did lead to the creation of the first intensity scale for earthquakes: the earliest, albeit crude, effort to quantify the phenomenon. The scale was the work of an Italian physician, Domenico Pignataro. He reviewed accounts of earthquakes in the whole of Italy – 1,181 in all – during the period from 1 January 1783 to 1 October 1786. Pignataro categorized the quakes according to their number of fatalities and their level of damage, as 'slight', 'moderate', 'strong' and 'very strong' – except for the six Calabrian earthquakes, which he judged to be 'violent'. It was a rough-and-ready beginning to earthquake measurement. Refinement had to await the occurrence of another devastating shock, in an area closer to Naples – the province of Basilicata – in mid-December 1857. With an estimated magnitude of 6.9–7.0 and as many as 19,000 fatalities, this was supposed to be the third greatest European earthquake ever recorded, after the Lisbon earthquake in 1755 and an earthquake at Catania in Sicily in 1693.

When the news from Naples reached Britain, it immediately attracted the attention of a brilliant Irish civil engineer, Robert Mallet, who was also a fellow of the Royal Society. Mallet had become interested in earthquakes back in 1830, after seeing a diagram in a book showing how the upper sections of two stone pillars in Calabria had been twisted by an earthquake, leaving the pillars still standing. Having attempted to explain the natural forces involved in such a distortion, without finding a satisfactory solution, Mallet was soon fascinated by earthquakes.

Over twenty years, he collected as much data about historical quakes as possible. His catalogue of world seismicity contained 6,831 listings, giving the date, location, number of shocks, probable direction and duration of the seismic waves, along with notes on related effects. From 1851 onwards, he also experimented with artificial earthquakes by exploding underground charges of gunpowder – as much as 5.5 tons of explosive in a single blast in a quarry in Anglesey in 1861 – thereby becoming the pioneer of this technique in exploration seismology, which is used by both scientists and industrial companies. He utilized an accurate stopwatch (designed by physicist Charles Wheatstone in 1840) to time the period between the explosion and the ripples created on the surface of a container of mercury. An illuminated image of cross-hairs was projected on the mercury and its reflection viewed through an eleven-power magnifier; the slight shaking in the mercury caused by the 'earthquake' made the reflected image blur or disappear. This proto-seismometer gave Mallet the speed of earthquake waves passing through different kinds of material: nearly twice as fast through granite (507 metres, or 555 yards, per second) as through sandy soil (304 metres, or 333 yards, per second), he calculated. However, the speeds were much lower than they should have been, indeed lower than Mallet himself expected (2,440 metres, or 2,668 yards, per second) – possibly because his mercury seismometer failed to detect the arrival of the earliest, and hence fastest, waves.

Less than a fortnight after the Italian earthquake on 16 December 1857, Mallet appealed to the Royal Society for a grant to cover part of the cost of examining the affected area of the kingdom of Naples. His letter from Dublin to the society's president in London argued that: 'The very recent occurrence of a great earthquake in the Neapolitan territory presents an opportunity of the highest interest and value for the advancement of this branch of Terrestrial Physics.'

He then coined a new term for what he had in mind:

> Within the last ten years only[,] Seismology has taken its place in cosmic science – and up to this time no earthquake has had its secondary or resultant phenomena sought for, observed, and discussed by a competent investigator – by one conversant with the dynamic laws of the hidden forces we are called upon to ascertain by means of the more or less permanent traces they have left, as phenomena, upon the shaken territory.

> Observed without such guiding light, or often passed by unnoticed and undiscovered for want of it, the facts hitherto recorded are in great part valueless – but with this guide such investigation is capable of results of high importance. Thus it was that [the] elaborate record of the effects of the great Calabrian earthquake [in 1783] is of so much less value than it might have been.[2]

But, Mallet now remarked, he could not afford to bear the entire cost of a field investigation in Italy. 'Were I a wealthy man I should proceed instantly, and on my own responsibility; but, although willing to give one-twelfth of my professional time and income for 1858 to it, private duties make it unsuitable that I should also be at the necessary expenditure for the journey and local inquiry.'[3]

The money required, £150, was promptly forthcoming from the Royal Society. Both the Geological Society of London and the Royal Geographical Society backed Mallet's proposal, as did the minister of foreign affairs in London, Lord Clarendon, and the archbishop of Westminster, Cardinal Wiseman (clearly a useful ally for a traveller in the interior of a strongly Catholic country). So did various scientists, beginning with the foremost geologist of his day, Charles Lyell.

Less than two months after the earthquake, Mallet was hard at work in Basilicata, travelling across the region on horseback. It was often tough going, with much rain and intense cold at night, staying in precarious lodgings, apart from a few monasteries and large houses. Even without the devastation of an earthquake, the region was not easy to penetrate. Much of it lay between 1,000 and 1,500 metres (3,300–5,000 feet) above sea level, with higher peaks. 'The settlements clung defensively to the lower summits, and the isolation of the region had in past centuries made it a refuge for many disparate peoples', write two recent commentators on Mallet's great investigation, Graziano Ferrari and Anita McConnell, in the Royal Society's *Notes and Records*. 'The only well-built and substantial buildings belonged to the religious communities and some minor gentry.'[4] Mallet, lacking provincial Italian, had to gather information through interpreters in a variety of dialects, apart from one direct conversation – in Latin! – with a padre.

'For a time the phenomenon seemed so complex – sometimes so contradictory apparently – that I began to fear I should get nothing but disjointed facts', Mallet wrote home to Lyell on 28 February 1858.

> [T]he aspect of some of the totally overthrown towns like Saponara is truly awful – there 3,000 persons at the least were *killed* in 30 seconds and the summit of the conoidal hill on which once stood a large and solidly built town is now but a rounded knoll of stones and rubbish out of which blackened beams stand up at every angle against the sky.[5]

In the village of Montemurro, he visited some ruins next to the tower of a fallen church, as he recalled much later in the tranquility of Dublin.

> [A]s I reached the level of the interior, something caught my foot – I stooped and found it was a long and broad piece of ancient-looking lace, that had decked some crushed altar. While I looked at it, a large piece of wall came toppling from the still standing part of the tower above, and crashed into fragments upon the talus of rubble down which I had just come, and over which I had again to make my way back to our headquarters.[6]

Out of a population of 7,000 people in Montemurro, 5,000 had died and 500 had been injured.

'At first sight, and even after cursory examination, all appears confusion. Houses seem to have been precipitated to the ground in every direction of azimuth. There seems no governing law, nor any indication of a prevailing direction of overturning force', Mallet wrote in his official report on the earthquake. This confusion made him extra-determined to avoid the mistake of the 1783 earthquake commission, which lacked knowledge of mechanical engineering. As he observed of his general approach:

> It is only by first gaining some commanding point, whence a general view over the whole field of ruin can be had, and observing its places of greatest and least destruction, and then by patient examination, compass in hand, of many details of overthrow, house by house and street by street, analysing each detail

> and comparing, as to the direction of force, that must have produced each particular fall, with those previously observed and compared, that we at length perceive, once [and] for all, that this apparent confusion is but superficial.[7]

For instance, the candlesticks in one particular church in the town of Moliterno, the Chiesa della Santa Dominica della Rosario, had fallen in two revealing ways:

> On altars, both at the north and south sides of the church, I found wood gilded candlesticks, that had been thrown out of plumb. Those on the north side had been thrown towards the N.W. at various angles, and still leaned against the back wall or shelf of the altar. They were high up and out of reach. Those on the south of the church, I found now in their usual places, but the Sacristan informed me, that all the wood candlesticks at that side, had been thrown quite off the altar, and were found scattered about the floor, and had been since replaced.
>
> These are decisive as to *direction* of the wave here, viz., from a point W. of north towards the south. The candlesticks at the north side, thrown by the first semiphase of the wave, were limited in motion, by the wall against which they fell and leaned, having gone over too far to recover their position at the return stroke. Those at the opposite, or south side, were at the same moment thrown out of plumb forwards, or towards the front of the altar, and finding no wall to support them, fell altogether.[8]

By assessing each crack of the earthquake's damage with a trained eye, Mallet compiled isoseismal maps: that is, maps with contours of equal earthquake damage/intensity (a method employed today, with refinements, to map seismic hazard). Although he placed too much reliance on the direction of fallen objects and the type of cracks in buildings as indicators of earthquake motion – cracking is in fact mainly a function of the type of building construction – Mallet's maps allowed him to estimate the centre of the shaking and the size of the earthquake relative to other earthquakes. However, he never stumbled on the ground rupture created by the geological fault that slipped in the 1857 quake. It was discovered only recently by geologist Lucilla Benedetti and colleagues following extensive field investigations, which revealed a hillside riven with ruptures. Why did Mallet miss these ruptures? One reason may have been that in 1857 – decades before great ruptures were observed in Japan in 1891 and California in 1906 – no geologist expected to see a surface rupture, because seismic movement was at this time assumed to take place deep underground. Another reason is that the ruptures may have been hidden by snow. 'Had he noted the 16-kilometre crack through the countryside, it seems very likely that this mechanically inclined individual would have made something of a link between fault rupture and quakes', comment seismologists Hough and Bilham.[9]

Nonetheless, 'Mallet's Report on the 1857 earthquake constitutes a watershed between descriptive naturalism and the scientific and speculative observation of an earthquake', Ferrari and McConnell conclude.[10] In it he used the new technique of photography, including stereoscopic images, to document the damage in detail. His subsequent lengthy text, with illustrations, was published in a two-volume study, *Great Neapolitan Earthquake of 1857: The First Principles of Observational Seismology*, after a frustrating delay, in 1862. He also published a catalogue of historical earthquakes that remains a standard reference today, and some extraordinary maps of seismic intensity throughout the world's landmasses that were not bettered until the seafloor measurements of the 1950s. Mallet's maps gave the first indication that earthquakes cluster in certain belts around the earth. 'Generally . . . the seismic foci, or the bands of earthquake disturbance in the Italian peninsula follow the lines of the great mountain ranges,' he wrote presciently, and went on to observe that: 'the same forces, wherever they may be, that develop themselves, as volcanic vents, and as earthquakes, are operative *everywhere* along the lines of the seismic bands; that is to say, along the axial lines of nearly all the great mountains ranges upon our globe.'[11] An explanation of why this is so, involving plate tectonics, would have to wait for another century, but in the meantime Mallet's map focused the attention of geologists on the mysterious global seismic intensity patterns.

The intensity of an earthquake is not be confused with its magnitude, the figure usually reported in newspapers. Intensity measures the size of a quake, as does magnitude, but whereas magnitude is calculated from the vibration of a pendulum in a seismograph, intensity is based on visible damage to structures built by humans; on changes in the earth's surface such as fissures; and on felt reports, for example, the effects during a quake on suspended light fittings or on a person driving a car. Intensity measures what human beings see and feel as a result of an earthquake; magnitude, what scientific instruments detect. Intensity is like the signal strength of a radio station, which depends on the listener's location and the radio waves' propagation path from the station to the listener; magnitude is like the power output of the station (measured in kilowatts). An earthquake can have only one magnitude, fundamentally; but it may have many intensities.

The particular intensity scale normally used today – there are several others – is a version of that created by Italian volcanologist Giuseppe Mercalli in 1902, modified in 1931. Running from intensity level I to XII, it has major drawbacks. In the first place, much of the scale is subjective, for example, its description of intensity level VII: 'Everybody runs outdoors. Damage negligible in buildings of good design and construction; slight to moderate in well-built ordinary structures; considerable in poorly built or badly designed structures; some chimneys broken. Noticed by persons driving motor cars.' Second, it depends on building construction quality, which cannot be easily assessed: for instance, one house may remain standing in an earthquake, while another house next door falls. Third, it is 'culturally' dependent: an intensity indicator useful in one context may be useless in another. Damage to stone and reinforced concrete buildings is important in, say, Tokyo, but scarcely relevant in an Indian village. In fact, Californian seismologists have suggested further modifying the Modified Mercalli scale for California so as to include the level of disturbance in grocery, liquor and furniture stores, and even the motion induced in waterbeds. Finally, and least satisfactorily, the Mercalli scale takes no account at all of the observer's distance from the epicentre: a small earthquake close by the observer can register a higher intensity than a large one far away from him.

Still, intensity scales are extremely useful. Many areas of the world lack seismographs capable of measuring the ground motion in strong earthquakes. Moreover, the seismic record before about the 1930s (when magnitude measurement was introduced by Richter) consists only of intensity reports. Intensity is thus the only quantitative way to compare a pre-20th-century earthquake with a modern one.

After Mallet's work came to an end in the 1870s, the deeper understanding of earthquakes, including the measurement of their magnitudes, required the development of the modern seismograph. The essential element in all of these (and presumably in the ancient Chinese seismometer of Zhang Heng) is a frame from which is freely suspended a mass – a pendulum – which either swings from side to side or oscillates up and down on a spring. The swinging mass responds to the transverse, horizontal motions of the ground; the oscillating mass to the vertical ground motions. The inertia of the suspended mass makes it lag behind, rather than coincide with, the movement of the frame (which naturally follows the shaking of the ground): the mechanism of the seismograph permits this relative motion of the mass and the ground to be recorded. If the movement of the mass is so arranged as to record a trace on a roll of uniformly moving paper, or nowadays a digital trace on a computer screen, then the seismometer is known as a seismograph.

The first seismometer intended to record the relative displacement of the ground and a pendulum bob was built in 1751 in Italy by Andrea Bina. However, it used a pointer attached to the swinging bob positioned so as to leave a trace in a static bed of sand; and there is no record of its actual use to measure an earthquake. Other seismometers, including Mallet's mercury-based instrument, followed during the next century. But the earliest modern seismograph – that is, an instrument capable of recording both horizontal and vertical motions with a timed trace on a moving surface – did not arrive until 1875. Invented by another Italian, P. F. Cecchi, it introduced the idea of two swinging pendulums to measure horizontal motions at right angles to each other and an oscillating pendulum on a spring to measure vertical motion. But it was insensitive and was scarcely used; the earliest surviving seismogram from a Cecchi seismograph dates only from 1887. During the 1880s, it was superseded by exciting seismographic developments in Japan. 'Cecchi's seismograph notwithstanding,' note two historians of early seismometry, James Dewey and Perry Byerly,

> it seems clear to us that the credit for the introduction of the seismograph in seismology belongs to a group of British professors teaching in Japan in the late

> 19th century. These scientists obtained the first known records of ground motion as a function of time. Furthermore, they knew what such records could reveal about the nature of earthquake motion. They used their instruments to study the propagation of seismic waves, and they used them to study, for engineering purposes, the behaviour of the ground in earthquakes.[12]

Here, a little historical background is required. After the visits to Japan of Commodore Perry in 1853 and 1854, the Ansei earthquake of 1855 and the restoration of the Meiji emperor in 1868, Japan rapidly opened itself to the West. The Japanese government hired foreign staff in scientific and technical positions in its new ministries and as teachers at the newly created Imperial College of Technology in Tokyo (which became part of the Tokyo Imperial University in 1886). Between 1865 and 1900, somewhere between 2,000 and 5,000 western experts – mostly in their twenties – went to work in Japan. Officially known as *oyatoi-gaikokujin* – which literally meant 'honourable foreign menial or hireling' – they were intended to help modernize Japan, without being given positions of real power that might relegate the strongly nationalistic Japanese to the status of colonial subjects.

Ironically, seismology was *not* one of the areas in which the foreign experts were invited to work. The Japanese were so acclimatized to the shaking of the earth that they felt no urgency to study earthquakes. Foreigners living in Japan were, however, acutely aware of the unfamiliar phenomenon. At the Imperial College, earthquakes were a constant subject of conversation among the expatriate teachers in the 1870s. 'We had earthquakes for breakfast, dinner, tea and supper', said one of them, John Milne.[13] As a result, seismology in Japan was founded not by Japanese but by foreign professors from related disciplines, such as geology, mining and civil engineering, the majority of whom came from Britain. Eventually, in 1886, the world's first chair in seismology was established at the Tokyo Imperial University, and filled by a protégé of Milne, Sekiya Seikei. Other notable Japanese seismologists, such as Fusakichi Omori, appeared before the turn of the century, when Japan was for a brief period the leading centre for seismology in the world, with its epicentre at the Tokyo Imperial University. (America had no teaching course in seismology until as late as 1911, five years after the great San Francisco earthquake, when a course was introduced at the University of California in the guise of geology.)

Milne was the most significant of these foreign professors in Japan. He came to Tokyo from England in 1876, at the age of twenty-six, as a professor of mining and geology; married a Japanese woman (the daughter of a Buddhist abbot) in the 1880s; and lived in Japan until 1895 – a period much longer than most *oyatoi-gaikokujin*. His professional work naturally required him to travel extensively within Japan, including some daring visits to volcanoes. In 1877, Milne and others chartered a special steamer to visit the erupting volcano of Oshima, a small island in Sagami Bay approximately to the south of Yokohama. He reported from the island that: 'Earthquakes, although so common on the mainland, are said not to occur here and the only shocks that have been felt are those which were produced at the time of the breaking out of the volcano.'[14] Much later, Milne concluded: 'In Japan, the majority of earthquakes which we experience do *not* come from volcanoes nor do they seem to have any direct connection with them. In the centre of Japan there are mountainous districts where active volcanoes are numerous, yet the area is singularly free from earthquakes.'[15]

In 1880, following a sharp earthquake in Tokyo–Yokohama, Milne urged the founding of the Seismological Society of Japan – the first body in the world devoted to seismology – headed, at Milne's insistence, by a Japanese. His seismograph, designed in 1881 with two other British scientists employed at the college in Tokyo, James Ewing and Thomas Gray, was widely deployed in Japan, beginning in 1883, when one was manufactured in Glasgow and presented to the Meiji emperor by the British Association for the Advancement of Science. After Milne returned to Britain in 1895, he was largely responsible for establishing a network of seismographic stations throughout the world, so as to collect data for evaluation at a central observatory, run by himself, located at his home on the south coast of Britain in Shide, on the Isle of Wight. This record was then published to an international audience in 'The Shide Circular Reports on Earthquakes', issued from 1900 until 1912, the year before Milne died. He also wrote a classic textbook on seismology. Although his theoretical contributions to the subject were small, Milne has a considerable claim to be the 'father' of seismology.

The Tokyo–Yokohama earthquake of 1880 opened up a split unique to Japan – between advocates of traditional Japanese buildings constructed in wood, and advocates of modern western architecture built in brick and stone. Japanese buildings were relatively immune to earthquakes, though prone to fire, whereas western-style buildings were prone to collapse in earthquakes, though relatively immune to fire. In the 1870s, after the annihilation of part of Tokyo in a fire, the modernizing government had constructed a brick-built residential and shopping area, a showpiece known as the Ginza, designed by an Irish architect; but despite being fire-proof, the Ginza had failed to attract residents and businesses, to the embarrassment of the government. Expensive rents, inconvenience and the unfamiliarity of the Ginza's layout and appearance played a part in this indifference; so too, it appears, did the fear of being trapped in a western-style building during an earthquake. When a new palace was designed for the emperor in a wholly western style, and construction began in 1879, it suffered from cracking during a minor earthquake and had to be modified. The ministerial offices in the new palace continued to be built in brick, whereas the residence of the emperor was switched to a wooden construction, designed by the official carpenter to the imperial court, rather than the original foreign architect. Clearly, there was more to this debate over Japanese versus western construction than simply engineering considerations. Milne soon found himself in the thick of it.

In 1880, he was keen to record the damage caused by the Tokyo–Yokohama earthquake and compare it with the damage recorded by Mallet after the Neapolitan earthquake in 1857. But few Japanese buildings in Tokyo demonstrated the data Milne required. With some regret, he observed:

> Everywhere the houses are built of wood and generally speaking are so flexible that although at the time of a shock they swing violently from side to side in a manner which would result in utter destruction to a house of brick or stone, when the shock is over, by the stiffness of their joints, they return to their original position, and leave no trace which gives us any definite information about the nature of the movement which has taken place.[16]

Indeed, some Japanese buildings, such as pagodas, employed elaborate wooden roof jointing explicitly to protect against earthquake damage.

The brick and stone buildings of recently constructed Yokohama, by happy contrast (at least for Milne), acted as 'one great seismometer'. Nearly all of their brick-built chimneys collapsed in 1880, along with a few of Yokohama's western-style houses – but none of its Japanese houses was much damaged, as in Tokyo, except for some warehouses that lost a little plaster. Therefore, in Yokohama, Milne sent out questionnaires to the foreign residents requesting them to report on the damage, for example, whether their windows had broken and at what time, and the direction of fall of their chimneys. By contrast, in Tokyo, he had to resort to examining the only bits of masonry that were widely available: gravestones in the countryside around the city, which had toppled during the earthquake in revealing ways. 'What consolation the residents in Yokohama may have received for the losses they have sustained, I am unable to say,' he wrote, 'but certainly if the houses in which they dwell had not been built our information about this earthquake would have been so small as to be almost valueless.'[17]

During the 1880s, Milne developed a system of monitoring earthquakes throughout Japan with the help of meteorologists, telegraph operators, the military and the state bureaucracy. Beginning at home, so to speak, he turned the Tokyo Imperial University's main building into a seismometer by tagging, dating and measuring existing cracks in its foundations, and attaching devices similar to seismographs directly to its walls. Then in 1882, he started to send out bundles of postcards to local government offices, such as post offices and schools, requesting a bureaucrat to mail a postcard to him in Tokyo every week recording the number of shocks he had felt. Having thereby established the areas of most frequent seismic activity, Milne then sited clocks in these areas that were designed to stop when sufficiently shaken, thus providing reports of the timing of earthquakes in different places that could be compared with each other. In 1883, precise instructions on how to keep these time records were issued to civil servants by the Seismological Society's Committee on a System of Earthquake Observations. Some of the clocks were placed in the field stations of the Imperial Meteorological Observatory. In the end, Milne 'successfully piggy-backed seismic monitoring onto the existing tasks of a host of government functionaries', notes historian Gregory Clancey in his pioneering study, *Earthquake Nation: The Cultural Politics of Japanese Seismicity*.[18]

In October 1891, Milne at last experienced a devastating Japanese earthquake to rival the one in Italy investigated by Mallet in 1857. It occurred in the alluvial Nobi Plain, known as 'the garden of Japan', and is today known as the Great Nobi earthquake or, more commonly, the Mino–Owari earthquake. Its shock waves were felt almost everywhere in Japan. Magnitude scales had not yet been invented, but on the evidence of contemporary seismograms, the earthquake's magnitude may have been as high as 8.4, making the Mino–Owari earthquake as large as the Great Kanto earthquake in 1923, maybe significantly larger. With about 7,300 fatalities, tens of thousands of injuries and over 100,000 people made homeless, this 1891 earthquake was the largest in Japan since the 1855 Ansei quake that destroyed Edo (Tokyo).

As for the damage, Milne reported in 1892 to the British Association for the Advancement of Science as follows:

> If we may judge from the contortions produced along lines of railway, the fissuring of the ground, the destruction of hundreds of miles of high embankments which guard the plains from river-floods, the utter ruin of structures of all descriptions, the sliding down of mountain sides and the toppling over of their peaks, the compression of valleys and other bewildering phenomena, we may confidently say that last year, on the morning of 28 October, Central Japan received as terrible a shaking as has ever been recorded in the history of seismology.[19]

The following year, 1893, Milne and a fellow engineer at the Tokyo Imperial University, William Burton, proved the truth of this report by publishing a book in Japan, *The Great Earthquake in Japan, 1891*, with a plate section of startling photographs of twisted railway lines, broken bridges and ruined factories. At the same time, Milne also joined the government's Imperial Earthquake Investigation Committee, formed as a direct result of the earthquake. He was the only foreigner to be invited to serve alongside Japanese engineers, architects, seismologists, geologists, mathematicians and physicists from the Tokyo Imperial University; and in due course he was honoured by the emperor with the Order of the Rising Sun. The committee was the first official Japanese organization to plan for the effects of future earthquakes.

Around this time, however, seismology took an international turn, with the invention of more sensitive seismographs in Europe, using optical, rather than mechanical recording, on a continuously moving photographic surface. From now on, earthquakes could increasingly be studied in seismological laboratories on the other side of the planet from their epicentres. In 1889, a seismograph located in Potsdam, Germany, designed by Ernst von Rebeur-Paschwitz, detected a large earthquake in Japan about an hour after it was felt in Japan. Milne confirmed the truth of this German observation. By 1893, inspired by these measurements in Europe, he had designed his most famous seismograph using photographic recording, which carried his name. Its sensitivity may have played a part in his decision to leave Japan and return to Britain in 1895, since he could now continue to study earthquakes in the islands of Japan from the Isle of Wight. (Another factor, however, is that Milne lost his entire house in Tokyo in a fire in 1895.)

America had been lagging behind Japan and Europe in seismology. But in 1886 it began to catch up, after an earthquake in Charleston, South Carolina. Felt over an area of 5 million square kilometres (2 million square miles), it was then considered to be the most destructive earthquake in the recorded history of the US, and is now allotted an estimated magnitude of 7.3.

Perhaps the Charleston earthquake's most important impact was on the United States Geological Survey, which was then concerned mainly with mapping and mining operations in the western part of the nation. The earthquake made geologists aware of the need for permanent seismographs. One of them, Joseph LeConte, wrote in the journal *Science* in 1887: 'we cannot any longer afford to study earthquakes without seismographs. The Geological Survey ought to have these in different parts of the country. The University of California has recently gotten three of best character (Ewing's and Gray-Milne's), which will soon be set up in different parts of the state.'[20]

Also important was a controversy that the Charleston earthquake provoked about the cause of earthquakes. According to the intuitively appealing 'shrinking earth' theory, 'an earthquake is a movement caused by a shrinking from a loss of heat, of the heated interior of the earth, and the crushing together and displacement of the rigid exterior as it accommodates itself to a contracting nucleus', asserted leading geologist John Strong Newberry in the *School of Mines Quarterly* of Columbia College, New York.[21] But a second geologist, Edward Everett Hayden, working with the USGS, immediately objected in *Science* to Newberry's assertion as follows:

> in view of the fact that many eminent scientific men are not prepared to subscribe to [this theory] at all . . . it is to be regretted that the author has not adopted the comprehensive and more non-committal definition given by Mallet . . . An earthquake is the passage of waves of elastic compression in the crust of the earth.[22]

Overall, writes a current US seismologist, Pradeep Talwani, the Charleston earthquake:

> led to a wider acceptance of Mallet's ideas about elastic strain, the nature of seismic waves, and a demise of the shrinking earth theory of earthquakes . . . The estimates of seismic velocity were the most robust at the time, and the effect of geological structures on the transmission of and attenuation of seismic waves was recognized.[23]

As for Japanese seismologists, they began to travel outside Japan, especially after Japan's victory in the Russo-Japanese War of 1904–5. In 1905, Omori went to India with a group of scientists and architects to record the aftershocks of a major earthquake in Assam. In 1906, he travelled to California after the famous earthquake, examined the ruins of San Francisco (where he was slightly injured by some stone-throwing, anti-Japanese, survivors) and wrote a report, which was the first detailed account of the earthquake to reach European scientists. A few months after the earthquake, a local newspaper, the *San Francisco Call*, carried a large photograph of

Professor Omori under the headline 'World's Greatest Seismologist Says San Francisco Is Safe'.[24] And in 1908, Omori investigated the appalling southern Italian earthquake in Messina, where 120,000 people died. There he concluded that the much smaller Japanese fatality figure in the 1891 Mino–Owari earthquake would undoubtedly have been far higher had ordinary Japanese homes been built of brick and stone, as in Italy, rather than of wood.

At around the same time, with the invention of increasingly sensitive seismographs, geologists and physicists, such as Lord Kelvin, began to understand the new potential of seismology for providing 'X-rays' of the inner structure of the earth, through calculating the speeds and trajectories of seismic waves as they passed via different regions of the crust, mantle and core. From the 1890s onwards, seismology would gradually be seen as part of the broad-ranging subject known as geophysics. Yet it still lacked – despite the pioneering field research of Mallet, Milne, Omori and many other scientists – any universal theory of where, when and why earthquakes occur, without which there could be no hope of forecasting their future occurrence. The great San Francisco earthquake would provide the first reliable clues to such a theory.

CHAPTER

6 ELASTIC REBOUND: SAN FRANCISCO, 1906

'Mole track' northwest of San Francisco thrown up by the earthquake on the San Andreas fault in 1906.

The most famous earthquake in history, which struck San Francisco in April 1906, produced a remarkable rupture of 476 kilometres (296 miles) in the surface of rural northern California from Monterey County in the south to Mendocino County in the north. The main shock – with an estimated magnitude of 7.7–7.9 – originated at a depth of nearly 10 kilometres (6 miles) under the sea about 3 kilometres (2 miles) west of the shoreline at Golden Gate Park, 13 kilometres (8 miles) west of downtown San Francisco. It tore up the ground like a giant mole, creating a furrow-like ridge about one-third of a metre (1 foot) in height and one metre (3 feet) in width. The waves propagated northwards and southwards at the rate of some 2.5 kilometres (1.5 miles) per second – ten times the cruising speed of a commercial jet airliner – as calculated from the global record of ninety-six seismographs; they reached the furthest of these instruments, on the island of Mauritius in the Indian Ocean, after eighty-four minutes, having travelled 12,600 kilometres (7,800 miles) through the earth. All along the mole-track's length, paths and roads, lines of trees and fences, and even whole barns, were offset up to 6 metres (20 feet) to the right (from whichever side of the rupture one looked at it). Along Bear Valley, near Point Reyes on the coast, towards the Mendocino end of the rupture, the ridge ran virtually straight for almost 25 kilometres (16 miles).

A famous folk tale about the earthquake – probably inspired by biblical stories of the earth's opening and swallowing cities – fooled a number of newspaper journalists, the president of Stanford University (which lost its new geology building in the quake) and even Grove Karl Gilbert, a leading geologist and member of the 1906 State Earthquake Investigation Commission. Gilbert reported that a cow had been swallowed by a fissure on a ranch in Olema, just south of Point Reyes Station, leaving only its tail, which was then eaten by dogs. He had not seen the cow or the protruding tail himself, and when he looked for a crack large enough to bury a cow, he failed to find it. Nevertheless, he said, 'the testimony on this point is beyond question'; the crack must have been caused by 'a temporary parting of the walls'.[1] In fact, it appears that a local rancher, Payne Shafter, with a dead cow to bury, had played a practical joke on Gilbert, whose credulity started the tale of the tail.

To explain how the rupture occurred, geophysicist Harry Fielding Reid – another member of the earthquake commission – came up with his model of 'elastic rebound' in 1906: the next major advance in seismology after the work of Mallet and Milne. Reid was the first scientist to perceive clearly that fault movement is the cause of earthquakes, and not vice versa. Although in practice Reid's mechanism suffers from many difficulties in describing how faults produce earthquakes, it is still the one most widely employed by seismologists in the 21st century.

Reid's theory starts with the concept of a geological fault as a joint between two rock planes, which can move both in a vertical direction (up and down) and in a horizontal direction (sideways). Since the joint is usually not exactly vertical, one plane overhangs the other. If the overhanging plane moves downwards, the fault is termed 'normal'; if it moves upwards, it is a 'reverse' fault. Movement in the vertical axis is known as 'dip-slip', that in the horizontal axis as 'strike-slip'. Of course, a real fault often shows both kinds of slip. Friction between the two planes of the fault determines its movement or lack of movement. The lower the friction, the weaker the fault and the more easily it slips. If the friction is low enough, the fault may slip constantly and aseismically; this is known as 'fault creep'. If it is of medium size, the fault may slip frequently, producing many small earthquakes. But if the friction is high, the fault may slip only occasionally, and there will be few, but large, earthquakes.

Well before 1906, geologists had been aware of a major fault in the area of the San Francisco earthquake. Indeed they had documented how roads, fences and streams crossing the fault area had become deformed by the movement of the fault. In 1895, it was named the 'San Andreas fault', after the small San Andreas reservoir located on the fault not far to the south of San Francisco, by geologist Andrew Lawson (who would be appointed chairman of the 1906 earthquake commission). But only in the decades after 1906 was the fault understood to be a great scar running up most of California, geologically highly complex, as much as 95 kilometres (60 miles) wide and 1,300 kilometres (800 miles) long. And only in the 1970s, with the advent of plate-tectonic theory, did it become clear that the San Andreas fault was the junction between two plates – the North American plate and the Pacific plate – grinding past each other in opposite directions.

Reid proposed that, prior to the 1906 earthquake, friction between the two planes of the San Andreas fault had locked part of the fault near San Francisco, deforming it during strike-slip movement of the planes – as a rubber band deforms when it is pulled apart. In fact, a current seismologist, Seth Stein, teaches 'elastic rebound' to undergraduate students using a rubber band. Stein attaches the band to a box containing a bar of soap and attempts to pull the box across a yoga mat. Pulling on the band at first causes the rubber to stretch elastically without moving the box, until the point when the forces acting in the band exceed the friction between the box and the mat – when the box suddenly jerks forward and the band snaps back to its original length. In a comparable manner, the San Andreas fault is imagined to deform until it exceeds the friction between its planes and then, during an earthquake, to snap horizontally as the planes spring away from each other and rebound into a less strained, offset, conformation, creating in the process a surface rupture. Thus a fault, after elastic rebound, recovers its original state before the build-up of strain (like the rubber band), yet is also reoriented (unlike the rubber band), since the two planes of the fault are now differently aligned with respect to each other.

'Elastic rebound' has resonance wider than seismology. It is a thought-provoking analogy for how San Francisco rebounded from its 1906 earthquake and fire, and reoriented itself to become one of the world's celebrated cities. And it applies to many other cities, too, which have rebounded after major earthquakes. 'These two words represent not only the single most fundamental tenet of earthquake theory but also the most apt metaphor to describe societal response to even the most catastrophic seismic events', claim seismologists Hough and Bilham in *After the Earth Quakes: Elastic Rebound on an Urban Planet*.[2]

The San Francisco earthquake occurred soon after five in the morning of 18 April 1906 in two shocks, separated by a pause of between ten and twelve seconds, lasting between forty-five and sixty seconds in all. The lower part of City Hall collapsed almost instantaneously, leaving a bare framework supporting a strangely intact cupola: one of the iconic images of the earthquake. A journalist from the *San Francisco Examiner*, who happened to be out early, recorded that as the dust cloud around City Hall cleared, 'The dome appeared like a huge birdcage against the morning dawn. The upper works of the entire building lay . . . in the street below.'[3]

At this implacable geological disjunction, police officer Edward Plume was sitting inside City Hall's police station. He later reported the terrifying experience:

> I was seated at a desk in the office when I felt a slight trembling of the great building. This had lasted a few seconds when Officer Jeremiah M. Dwyer said, 'that's an earthquake.'
>
> Then the seemingly light temblor [tremor] began to increase in violence and all of a sudden I was thrown clear out of the chair. As I was then being tossed from one side to the other by the shocks, I tried to find something to hold onto and prevent myself from falling.
>
> Meanwhile the noise from the outside became deafening. I could hear the massive pillars that upheld the cornices and the cupola of the City Hall go cracking, with reports like cannon, then falling with crashes like thunder. Huge stones and lumps of masonry came crashing down outside our doors; the large chandelier swung to and fro, then fell from the ceiling with a bang.
>
> In an instant the room was full of dust as well as soot and smoke from the fireplace. It seemed to be reeling like the cabin of a ship in a gale. Feeling sure that the building could never survive such shocks, and expecting every moment to be buried under a mass of ruins, I shouted to Officer Dwyer to get out.
>
> The lights were then out, and though the dawn had come outside, the station, owing to the dust and smoke inside, and the ruins and dust outside, was all in darkness.
>
> Dwyer and I made a rush for the nearest door, stumbling over chairs and desks and other litter as we scrambled out to Larkin Street. It was dark in the street, and choking with dust. We ran across a small alley. The dust from the buildings that were still falling made it impossible to see anything.
>
> As I reached the alley, the front walls of the Strathmore [apartment building] began to fall. There was a vacant lot about 50 feet away, and I ran there and stood along a high fence waiting for the ruin to become complete. Dwyer had tripped when crossing Larkin Street and had not joined me.
>
> Then the quaking stopped, and we could hear screams in all directions.[4]

Both Plume and Dwyer were extremely lucky to escape with their lives. In fact, only one on-duty police officer died as a result of the 1906 quake, when a wall fell on him.

Six blocks from City Hall, the supposedly earthquake-proof Palace Hotel 'seemed to dance a jig', according to a guest, the manager of a visiting opera company that had staged *Carmen* at the Mission Opera House a few hours earlier.[5] One of the singers, Enrico Caruso, the most famous operatic tenor in the world, felt that:

> Everything in the room was going round and round. The chandelier was trying to touch the ceiling, and the chairs were all chasing each other. Crash – crash – crash! It was a terrible scene. Everywhere the walls were falling and clouds of yellow dust were rising. The earth was still quaking. My God! I thought it would never stop.[6]

Having escaped from the Palace, Caruso walked to another hotel, the St Francis, in search of other members of the opera company. He was spotted near this hotel's entrance by a local photographer, wearing a fur coat over his pyjamas, smoking a cigarette and muttering ''ell of a place! 'ell of a place! I never come back here.'[7]

Less predictably, many others were thrilled by the earthquake. A second famous visitor to California, psychologist William James (brother of the novelist Henry James), had never experienced an earthquake before because he lived on the east coast. Staying in a flat on the campus of Stanford University, he awoke to find his bed vibrating, furniture crashing over and an awful roaring noise. But James did not hold with the Victorian belief in repression of strong emotions. He wrote soon afterwards in an article, 'On Some Mental Effects of the Earthquake':

> [My] emotion consisted wholly of glee and admiration . . . I felt no trace whatever of fear; it was pure delight and welcome.
>
> '*Go* it,' I almost cried aloud, 'and go it *stronger*!'
>
> I ran into my wife's room, and found that she, although awakened from sound sleep, had felt no fear, either. Of all the persons whom I later

> interrogated, very few had felt any fear while the shaking lasted, although many had had a 'turn', as they realised their narrow escapes from bookcases or bricks from chimney-breasts falling on their beds and pillows an instant after they had left them.[8]

In general, despite severe damage to Stanford's buildings, including the men's dormitory, two deaths and some serious injuries, 'Everybody was excited, but the excitement at first, at any rate, seemed to be almost joyous', James maintained. 'Here at last was a *real* earthquake after so many years of harmless waggle! Above all, there was an irresistible desire to talk about it, and exchange experiences.' For several subsequent nights, most people slept outdoors, partly for safety in case of a recurrence, 'but also to work off their emotion, and get the full unusualness out of the experience', thought James. 'The vocal babble of early-waking girls and boys from the gardens of the campus, mingling with the birds' songs and the exquisite weather, was for three or four days a delightful sunrise phenomenon.'[9] Apparently, James was not alone in his enthusiasm for the emotions roused by the destruction.

Others were not as fortunate as Plume, Caruso and James. In 1906, about one-sixth of San Francisco's population of 410,000 was living on 'made ground', that is land filled in by real-estate speculators in the second half of the 19th century. Made ground was extremely vulnerable to earthquake shaking. In the Sixth Street area, for instance, hotels for transients had been built on an infilled depression, once known as Mission Swamp, created by the city's first major earthquake in 1868. These hotels collapsed in a domino-like effect: Nevada House at 132 Sixth slammed into the Lormor at 136 Sixth, which fell upon the Ohio House at 142 Sixth, which in turn rammed the Brunswick House into Howard Street. The four hotels, combined, contained about a thousand rooms.

Another hotel, on Valencia Street, also built on a swamp, fared still worse. It sank three storeys and then collapsed upon itself, leaving just one storey above ground level. At the same time, the settling swamp broke cable car conduits and the major water main that supplied water to downtown San Francisco. Most of those trapped in the lower three floors of the Valencia Street hotel probably drowned within a few minutes of the earthquake.

In Chinatown, the damage was horrendous. Moreover, it revealed barricaded iron doors and false walls, constructed to hide gambling halls, opium dens and brothels. Two police officers who went to the whorehouses on nearby Commercial Street found the female prostitutes and their male customers running around naked. Many of the prostitutes were too scared to go into the ruins to find clothes, so the officers had to rummage on their behalf for whatever skirts and petticoats they could find. Most of the customers bolted as soon as they could, wearing whatever little they had on. Astonishingly, there is no record of any Chinese person being taken to an emergency hospital after the quake.

At Fire Station No. 6, the earthquake caused the rear of the two-storey wood-frame station to subside about a metre (3 feet) into made ground and the floor to part in the centre. The doors of the engine house shook open and the horses ran off into the streets. Only with great difficulty could the fire engine be got out of the station. It would very soon be desperately needed.

The fire started by the earthquake was by no means the first major fire in the history of San Francisco. During the Gold Rush period, fires burnt large parts of the city to the ground six times between Christmas Eve 1849 and June 1851, when wooden buildings were replaced by brick and stone ones in the business district. In 1850, the citizens emblazoned a phoenix on the city's official seal. However, the 1906 fire was by far the most destructive of these conflagrations. It burnt for three days until it was at last deprived of fuel around seven in the morning on 21 April. During this period, the fire devoured 508 blocks and more than 28,000 buildings, extending over 12.2 square kilometres (4.7 square miles) – three-quarters of the city (and about eight times the area destroyed by the Great Fire of London) – and cost at least $500 million. Combined with the earthquake, it left 225,000 people – more than half of the city's population – homeless.

San Francisco's fire department, crucially assisted by men from the port's naval base, was hamstrung by the failure of the main water supply as a result of earthquake damage to pipes, exacerbated by the destruction of the police force's telephone system and the fire department's telegraph system. (The postal system, astonishingly, continued to operate and lost not a single piece of mail in the fire, as a result of postal workers' beating out the flames with wet mailbags.) Eventually, the civil authorities were forced to give the military authorities permission to begin dynamiting and burning unburnt streets, so as to create fire-breaks. Block by block, army crews demolished houses, mansions and churches, which were immediately torched with kerosene-soaked rags hurled into the buildings. How far ahead of the fire to demolish buildings was a challenging problem, complicated by disagreement between the civil and military authorities and the fact that some of the targeted buildings belonged to political supporters of the city's notably corrupt mayor, Eugene Schmitz. He compromised and instructed that buildings should be dynamited only when they were 'about to burn'.[10] The fire continued to advance. When it reached Van Ness Avenue, there was no alternative but to blow up buildings well ahead of the inferno and create a burnt area too broad for the flames to bridge. During the final hours of the exhausting battle:

> Delirious fire-fighters collapsed in the street, only to be dragged from danger by refugees. Others rolled in the gutters to keep melted rubber turnout clothes from sticking to their bodies. From somewhere, said Captain Stephen Russell of Fire Engine no. 27, a doctor or nurse moved along the fire line, injecting strychnine, an alkaloid poison used for killing vermin, into the fire-fighters in the belief that a mild dose would ease the pain of burns and act as a stimulant.[11]

The description is from *Denial of Disaster*, a definitive joint study of the earthquake published in 1989 by Gladys Hansen, the archivist for the City and County of San Francisco, and Emmet Condon, a former fire chief of San Francisco who served as a fire-fighter there for over thirty years. Their title refers to San Francisco's long denial of the true number of fatalities from its 1906 earthquake and fire. For decades, the US Army's estimate of 498 deaths in the city was accepted, based on the body count in the ruins, the numbers of people reported missing and a small number of men executed for criminal acts. But in the 1960s, Hansen began to doubt this official figure when she was put in charge of the genealogy collection of the city's library. There she was regularly asked for a list of those who had died in 1906 – and discovered that no such list existed.

Two reasons made it difficult to reconstruct such a list. First, there were hundreds of unrecorded tragedies that occurred during the earthquake and fire. For example, 'drunks who refused to be pulled from water-front dives, complete families suffocated in ramshackle flats, infants who were never found or were killed in baby buggies by flying cinders', write Hansen and Condon.[12] Second, there were the indirect casualties: those who died well after April 1906, in the months following the earthquake, from their injuries, from illnesses contracted in the insanitary conditions prevailing in refugee camps after the disaster, from release of toxic materials and from psychological trauma, including suicide. After extensive research, Hansen concluded that the death toll from the earthquake and fire – if defined to cover all fatalities that occurred during the earthquake and fire plus any earthquake/fire-caused injury or illness that became fatal within a period of one year following the earthquake – was very much higher than the reported number. Her final figure came to more than 3,000 deaths, that is, six times the army's original figure. Almost a century after the earthquake, this more truthful estimate became widely accepted.

The denial of these deaths was part of a far broader picture of denial after the earthquake. One is reminded of the general underestimation of earthquakes by archaeologists and historians. In San Francisco, politicians and businessmen unquestionably made a sustained effort to blame the devastation on the fire rather than the earthquake, and this belief became firmly established – so much so that in later years people spoke of 'before The Fire' and 'after The Fire' as historical benchmarks. The belief found a ready response among ordinary San Franciscans for the following reasons. In the first place, building insurance policies generally covered fires, but not earthquakes, which naturally encouraged policy-holders to downplay the importance of the earthquake versus the fire. Second, focusing on the fire deflected attention away from any long-term seismic risk to the city, which was good for business. Third, blaming the fire encouraged everyone to rebuild the city as it was before, as quickly as possible, without the need for any expensive changes to the foundations of buildings or any anti-seismic structural engineering. San Francisco's daily newspapers abetted the belief by publishing telegraphic reports of small tremors in the eastern US, while omitting to report the stronger aftershocks in San Francisco itself.

Indeed, the State Earthquake Investigation Commission received little help from San Franciscans in gathering information. Pressure was exerted on its members by the city authorities and the chamber of commerce to suppress its findings, as had happened after the city's previous earthquake in 1868, when a disturbing scientific report was never published. One commission member in 1906, John Casper Branner, from Stanford University, candidly recalled in 1913: 'We were advised and even urged over and over again to gather no such information, and above all not to publish it. "Forget it," "the less said, the sooner mended," and "there hasn't been any earthquake," were the sentiments we heard on all sides.'[13] Another member, the geologist Gilbert, complained in his presidential address to the American Association of Geographers three years after the earthquake:

> The policy of assumed indifference, which is probably not shared by any other earthquake district in the world, has continued to the present time and is accompanied by a policy of concealment. It is feared that if the ground of California has a reputation for instability, the flow of immigration will be checked, capital will go elsewhere, and business activity will be impaired.[14]

Thus, a mere week after the earthquake, there was a meeting of the San Francisco Real Estate Board, reported in the *San Francisco Chronicle*. It passed a resolution that 'the great earthquake' was a phrase that should be replaced by 'the great fire'.[15] By implication, the buildings of San Francisco could rise again, phoenix-like, from the ashes of great fires, as they had done from previous fires; but there was no such assurance from the destruction caused by great earthquakes.

In the same week, an impromptu mass meeting assembled aboard a sleeping car in a train heading for the ruins of San Francisco from Chicago (which had suffered a great fire in 1871). One of those present, the head of the Chicago Bureau of Charities, later recalled:

> After half an hour of vigorous speeches, a resolution was adopted to great cheering and entire unanimity, announcing as a fact beyond dispute that the disaster in San Francisco was due solely to fire, to such a calamity, in short, as

> might occur in any well-ordered city, and the slight tremor which preceded the fire had nothing to do with the tragedy beyond, perhaps, breaking gas mains or water mains here or there.[16]

However, the most influential voice in support of this view was the biggest business in California: the Southern Pacific railroad company. Its rail facilities in northern and central California had sustained an enormous amount of earthquake damage, and various of its properties in downtown San Francisco had been lost in the fire. The company was also expanding at great cost in southern California. It therefore had a lot to lose if the state was projected to be at future risk from major earthquakes.

The company's general passenger agent wrote to chambers of commerce throughout California in order to 'set the record straight' – by refashioning the disaster. 'We do not believe in advertising the earthquake,' he explained. 'The real calamity in San Francisco was undoubtedly the fire.' He asked the chambers to assure the local press that violent earthquakes are 'so infrequent that no city in the temperate zone has ever been twice affected in a serious way by an earthquake'.[17] The chambers might also draw attention to the fact that San Francisco was principally a wooden city (except for the buildings of the business district, which largely escaped the fire) because of the cheapness of local timber, and hence the city was liable to destruction by fire. Furthermore, the chambers might explain that the city's water supply, though old and liable to subsidence in swamp land, was now going to be replaced by a system of reservoirs. Indeed, recommended Southern Pacific, the chambers of commerce should encourage lecturers on the disaster to dwell upon the rapidity and inventiveness of the city's recovery as reconstruction proceeded.

'The scope of the Southern Pacific Company's reworking of the history of the catastrophe was, and is, breathtaking', write Hansen and Condon. 'Continued use of this large collection of well-written and well-packaged data by authors and researchers has effectively placed [its] sanitized, simplistic and, in many cases, grossly inaccurate version of the earthquake's effects into nearly all subsequent books and articles about the San Francisco earthquake and fire.'[18]

Even the insurance companies in due course connived in the view that the disaster was pyrogenic, rather than seismic, in origin – despite a guarantee that it was against their financial interests to do so. A standard clause in fire insurance contracts throughout the US, the 'fallen building' clause, stated: 'If a building, or any part thereof, fall except as the result of fire all insurance by this policy on such building or its contents shall immediately cease.'[19] A building that fell in an earthquake, or from dynamiting, and which then caught fire, was not covered by such insurance.

Policy-holders were well aware of this. During the conflagration, rumours circulated about the implications, and some house-owners reacted by deliberately setting fire to their quake-damaged houses in the hope of collecting on their insurance. After the conflagration, the chairman of the Fire Underwriters' Adjustment Committee said: 'it may be considered an extraordinary thing that in the first 2,000 claims submitted to insurance men in San Francisco, after the earthquake and fire, every man filing a claim swore that his property was uninjured by the earthquake. I say, that is extraordinary.'[20]

The problem was, of course, how to distinguish in the ruins between the different kinds of damage caused by the earthquake, the fire and the dynamiting. There was much photographic evidence available (as witness the detailed images in *Denial of Disaster*), but also much room for dispute. In the end, an across-the-board compromise was reached, known as the 'horizontal cut', that is, a percentage discount of the full value of the insurance policy, allowing for the possible effects of earthquake or dynamite damage. Insurance adjusters pushed for the discount to be one-third, but after negotiations between the insurance companies and a committee of policy-holders, the discount came down to one-tenth. 'This discount was the basis for future statistics which attributed 10 per cent of the overall damage to the earthquake and 90 per cent to the fire', note Hansen and Condon.[21] For sure, it was a split of which the California boosters of the Southern Pacific Company must have approved.

Even so, most claimants lost out heavily because they had too little insurance. Of the 90,000 claims by individuals and companies, by far the largest was that of the Palace Hotel, where Caruso had stayed. It received the full value of its policies, $1,518,500, but its full loss, according to the hotel's manager, was actually $6 million. In due course, the building was razed, and its 31 million bricks carted away to use as land-fill. The total loss in the city from the earthquake and fire has been variously estimated as between $500 million and $1 billion – excluding losses outside the fire zone and the replacement or repair of the urban infrastructure (water and sewage systems, schools, fire houses, police stations and so on), plus the huge cost of the relief operation for the homeless. The cost of rebuilding the infrastructure – at any rate the proportion of it attributable to earthquake damage – was mostly financed by municipal bonds, some of which were not paid off until the 1980s, not long before the next major earthquake.

Whatever may have been the true total cost, it was large enough to produce both national and international financial repercussions. It was equivalent to between 1.3 and 1.8 per cent of the nominal US gross national product in 1906. On the New York stock exchange, there was some heavy selling immediately after the earthquake and fire. Yet, within a mere three days, stock prices were moving upwards again. On 23 April, a writer in the *New York Times* who had researched previous 'catastrophe markets' remarked, to his surprise, that such rapid stock price recovery was the rule rather than the exception after disasters.[22] Soon, San Francisco started to look like a promising place to invest, especially in the field of construction. 'In the logic of capitalism, destruction was conceived as an opportunity for modernization', notes historian Deborah Coen in *The Earthquake Observers*.[23]

Not until the autumn of 1906, when foreign insurers had to pay out on claims against their San Francisco policies, did the real financial turbulence begin. The hardest hit were British insurance companies, who were liable for an estimated $108 million, mainly in gold: the preferred form of payment in California since the days of the Gold Rush. Over a two-month period, the outflow of British funds represented a 14-per-cent loss in the gold money stock of England: the largest drop recorded in the period 1900–13. The ratio of reserves to deposits fell to its lowest level since the financial crisis of 1893. To stem the outflow of gold, the Bank of England began to discriminate against American finance bills and, along with other European central banks, to raise its interest rates, although it did not officially attribute the funds outflow to the San Francisco insurance liabilities. This European policy in 1906–7 helped to cause a short, but severe, US recession in May 1907, followed by a crash in the New York stock market in October 1907, and a financial panic. 'Responding to the panic, Congress passed the Aldrich-Vreeland Act in 1908; this created the National Monetary Commission and authorized banks to issue emergency currency backed by commercial paper in times of crisis', note economic historians Kerry Odell and Marc Weidenmier. 'The National Monetary Commission's recommendations eventually formed the basis of the Federal Reserve System that was established in 1913.'[24] Thus, the San Francisco earthquake of 1906 might be said to have triggered the formation of the US central bank.

Along with the insurance adjusters came an army of researchers from engineering societies, universities and government agencies. They stayed for longer, about three years, treating downtown San Francisco as a laboratory for studying the structural engineering aspects of the earthquake and fire – like Mallet in the ruins of the great Neapolitan earthquake half a century earlier. Their aim was to discriminate between damage caused by the earthquake, dynamiting damage and fire damage, and thereby to provide a basis for rebuilding a more earthquake- and fire-resistant city.

But it was not to be. An engineering report published soon after the earthquake predicted: 'It is probable that the new San Francisco to rise on the ruins will be to a large extent a duplicate of the former city in defects of construction.'[25] Decades later, in *Denial of Disaster*, Condon predicted: 'the hazard of fire to the city after a similar disaster today is much greater than in 1906.'[26] In 1989, during the third major earthquake to strike San Francisco (magnitude 6.9), although the city fortunately avoided a great fire, it once again experienced major seismic damage, particularly on made ground. In 1991, the *Bulletin of the Seismological Society of America* (a scientific association founded in 1906) commented in an 800-page special report on the 1989 earthquake:

> The effects of earthquake strong ground motion on unreinforced masonry, soft first stories, decayed timbers, bad foundations, hydraulic fill and young Bay mud hardly qualify as news, especially in San Francisco where these 'lessons'

had all been learned in 1906 if not before . . . Until there is a permanent, national consciousness that the hazards from earthquakes are very real and the potential losses very great, it seems inevitable that we shall learn the lessons of 1906 and 1989 yet again.[27]

In other words, San Francisco rebounded after 1906, but retained most of its earlier faults. The gleaming new city that emerged from the ashes of the old city was structurally less sound than the pre-1906 city. In 1915, it was showcased to the world at the Panama-Pacific International Exposition, which was built on land reclaimed from San Francisco Bay. After this had closed, the site was developed as the city's Marina district. In the 1989 earthquake, the Marina district suffered the most severe damage in the city. Bits of charred wood and pieces of imitation travertine – presumably from the fire in 1906 and the broken-up debris of the 1915 exposition – were ejected in sand boils when the made ground in the Marina liquefied.

Not until 1933 did the state of California formally begin to protect its citizens against earthquake destruction. In that year, a quake struck Long Beach near Los Angeles, killed 120 people and caused property damage estimated at $50 million. Only the lateness of the hour – just before six in the afternoon – saved hundreds of schoolchildren from almost certain death. Within a month, California introduced strict design and construction regulations for public schools.

Many residents somehow convinced themselves that the Long Beach earthquake, magnitude 6.4, was southern California's equivalent of the great earthquake in northern California in 1906. 'Though only a moderate earthquake', wrote Charles Richter in 1958, it was regarded as a 'major disaster'. But at least, he added, 'This calamity had a number of good consequences. It put an end to efforts by incompletely informed or otherwise misguided interests to deny or hush up the existence of serious earthquake risk in the Los Angeles metropolitan area.'[28]

A mixture of almost total ignorance about earthquakes and false optimism about human beings prevailed among the Californian public at this time. Journalist and social critic Carey McWilliams collected together some of the folk tales circulating in the local newspapers around Long Beach immediately after the 1933 quake:

> That an automobile, while being driven along a boulevard in Long Beach, shook so hard that it lost all four tires; that the undertakers in Long Beach didn't charge a penny for the 60 or more interments following the quake; that the quake was the first manifestation of the awful curse placed on southern California by the Rev. Robert P. Shuler after its residents failed to elect him United States Senator; that sailors on vessels a mile or more off shore from Palos Verdes saw the hills (quite high) disappear from sight; that the bootleggers of Long Beach saved hundreds of lives by their public-spirited donation of large quantities of alcohol to the medical authorities; that women showed the most courage during the quake and that men can't stand up under earthquakes; that the shock of the quake caused dozens of miscarriages in Long Beach, and that an earthquake will often cause permanent, and annoying, irregularities among women; that every building in southern California that was not damaged by the quake is 'earthquake proof' . . . [and finally] that the earthquake, followed by the appearance of a mighty meteor on 24 March, presages the beginning of the end.[29]

Not until 1948, after much debate, did the state of California impose minimal seismic design requirements on buildings other than schools. And not until 1990, after the third major earthquake in San Francisco, did the state government issue a detailed plan of what should happen in the event of an earthquake warning.

Meanwhile, insurance companies prospered, as anxious institutions and householders who had previously avoided taking out earthquake insurance finally signed up for it. Stanford University – which had been dramatically damaged in 1906 – took earthquake insurance in 1980, but dropped it in 1985 after being offered a policy with a premium of $3 million per year, a $100 million deductible (excess), and coverage for only $125 million loss above the deductible. In the 1989 earthquake, without any insurance, the university suffered a loss of some $160 million.

Yet even those with the most knowledge of earthquakes – seismologists – are divided as to whether earthquake insurance is worth the high premium and enormous deductible. In the 1990s, despite several recent earthquakes in northern and southern California, the chairman of the California Seismic Safety Commission and a worldwide expert on seismic hazard, Lloyd Cluff, did not bother with earthquake insurance on his older house in San Francisco (built by the person who was the engineer for the Golden Gate Bridge), because the amount of the deductible made the insurance pointless. Nor did the director of the seismology laboratory at the California Institute of Technology and one of the world's leading seismologists, Tokyo-born Hiroo Kanamori, in regard to his house in Pasadena, not far from Los Angeles. According to Kanamori, 'My personal way of dealing with earthquakes is that I want to reduce the worry I have. I live in a relatively inexpensive home, a cheap house, really, so that the loss will be limited. I am very happy about it.'[30] Not only did Kanamori carry no earthquake insurance; his house was not even retrofitted against earthquakes, unlike many older houses and other buildings in California.

The Californian attitude to earthquakes has always been an ambivalent mixture of embrace and denial. Only in Japan is the protection of urban infrastructure, buildings and people against earthquakes treated with deadly seriousness. But then, Japan is the only industrialized nation in the world ever to have suffered the almost total destruction of its capital city by an earthquake.

CHAPTER

7 HOLOCAUST IN JAPAN: TOKYO AND YOKOHAMA, 1923

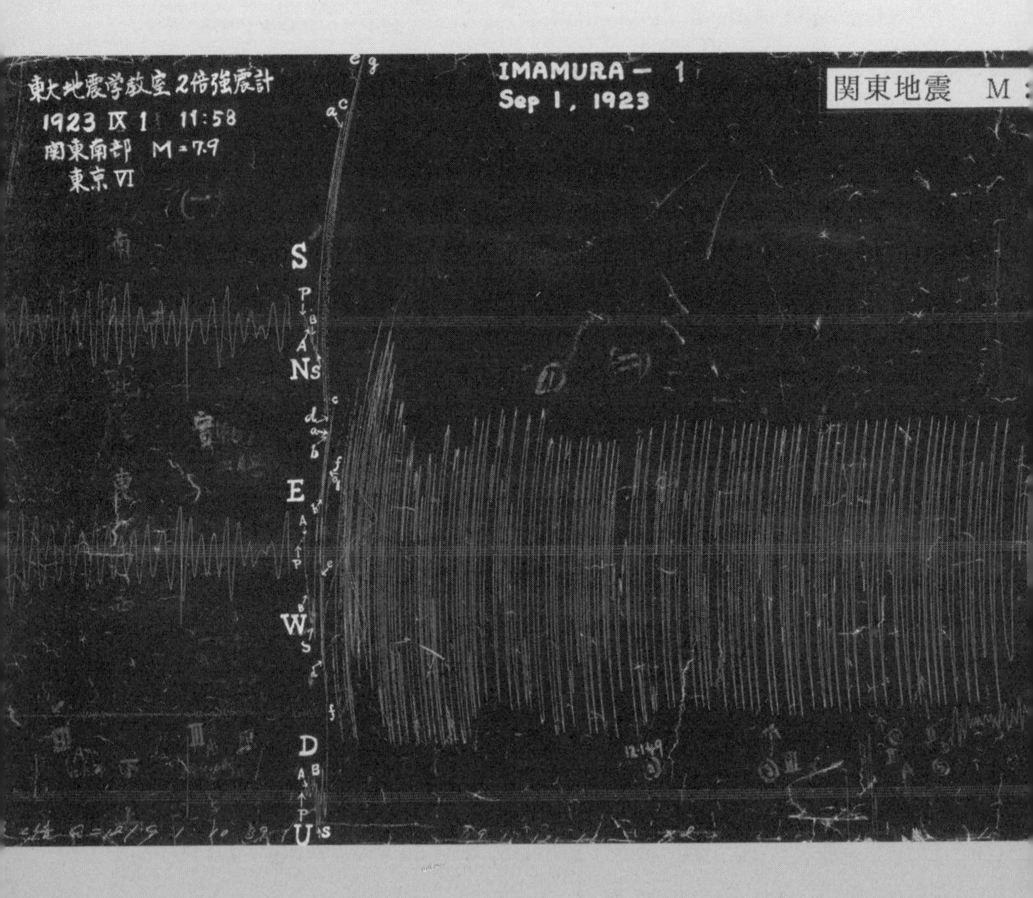

Seismogram recorded at Tokyo's Imperial University by Akitsune Imamura on 1 September 1923.

The Great Kanto earthquake, Japan's worst ever earthquake – which took at least 140,000 lives in its subsequent fires – struck Tokyo and surrounding areas of the Kanto region just before lunchtime on 1 September 1923. The capital city was subjected to between four and five minutes of shaking (accounts differ), followed shortly after by a tsunami 11 metres (36 feet) in height. The energy released was equivalent to some 400 Hiroshima-sized atomic bombs.

Charcoal and gas braziers were then cooking the midday meal in a million wooden houses. Soon multiple small fires started in panic-stricken kitchens and, feeding on the congested houses, merged to form terrifying firestorms that burnt through the night. By the morning of 3 September, 18 square kilometres (7 square miles) of Tokyo – two-thirds of the city – had been incinerated (as compared with 12.2 square kilometres, or 4.7 square miles, of San Francisco in 1906).

Nearby Yokohama, the international port of Tokyo, with its preponderance of recent stone and brick structures built in imitation of a European city, suffered more from the shaking than Tokyo, before it, too, succumbed to fire. A passenger on board a ship in Yokohama's harbour reported that about a minute after the first shock, a yellow cloud – 'very thin at first but growing in size every second' – rose up from the land. It 'formed a continuous strip all around the bay . . . deepening in colour, travelling at great speed toward the north. This cloud was doubtless caused by the dust from collapsing buildings'.[1]

All that remained of the city's Grand Hotel were piles of charred rubble. Even now, in the small sepia-toned photograph printed in the official 1926 report on the earthquake, introduced by Japan's prince regent (the future Emperor Hirohito), the utterly ruined hotel is an arresting sight. The photo's caption states baldly: 'No human work can withstand the violence of Nature.'[2]

A Tokyo schoolboy of thirteen, who later became Japan's most famous film director, Akira Kurosawa – legendary for his dramatization of the extremes of human behaviour in such movies as *Rashomon*, *Seven Samurai* and *Throne of Blood* (a version of *Macbeth*) – was then living in a hilly suburb. His family house was badly damaged and its electricity supply knocked out along with the power in the rest of the city, but he and his family were lucky to escape physically unscathed.

In his wise and haunting autobiography, written nearly six decades later, Kurosawa devotes three sections of the book – 'September 1, 1923', 'Darkness and Humanity' and 'A Horrifying Excursion' – to his teenage experience of the Great Kanto earthquake. He observes: 'Through it I learnt not only of the extraordinary powers of nature, but of extraordinary things that lie in human hearts.'[3]

As Kurosawa notes, following the great fire, a rumour swept through the darkened city that Korean immigrant labourers – a group generally despised by the native Japanese as a result of Japan's annexation of Korea in 1910 – had started the conflagration. The Koreans were also said to be poisoning Tokyo's wells; and even, according to Japanese-owned newspapers, to be plotting the assassination of Japan's imperial family.

In the ten weeks or so after the earthquake and fire, during which martial law was imposed by the government, between 6,000 and 10,000 Koreans – out of perhaps 12,000 Koreans then working in the Tokyo–Yokohama area – were lynched by Japanese vigilantes, some of them with the connivance of nationalistic members of the military and police; the true number of victims is unlikely ever to be established, in the absence of any official enquiry since 1923 (or any formal apology for the massacre by the Japanese government). A leading seismologist, Akitsune Imamura, narrowly escaped being murdered while cycling home late at night, because of his strange-looking helmet. Even the car of the newly appointed Japanese prime minister was nearly attacked while he was returning home from his first cabinet meeting, on the night of 2 September, by vigilantes led by a man wielding a large club, before the thugs became aware of the identity of its shocked occupant.

Kurosawa's military-minded father – something of a samurai to his son – while going in search of missing relatives in a burnt-out area, was mistaken for a foreigner because of his full beard and surrounded by a mob wielding clubs, who dispersed only when he thundered 'Idiots!' at them in Japanese At home, the young Akira was told to keep watch at night, with a wooden sword in his hand, over a drainage pipe, narrow enough for a crawling cat, in case Koreans were able to sneak through it. He was warned, too, not to drink the water from a neighbourhood well because its surrounding wall carried white chalk marks written in a strange Korean code. But the grimly absurd truth was that Akira himself had been responsible for these meaningless scribbles.

When the holocaust in central Tokyo had died down, Kurosawa recalls how his elder brother invited him to take a look at the ruins:

> I set out to accompany my brother with the kind of cheerfulness you feel on a school excursion. By the time I realized how horrifying this excursion would be and tried to shrink back from it, it was already too late. My brother ignored my hesitation and dragged me along . . .
>
> At first we saw only an occasional burned body, but as we drew closer to the downtown area, the numbers increased. But my brother took me by the hand and walked on with determination. The burned landscape for as far as the eye could see had a brownish red colour . . . Amid this expanse of nauseating redness lay every kind of corpse imaginable. I saw corpses charred black, half-burned corpses, corpses in gutters, corpses floating in rivers, corpses piled up on bridges, corpses blocking off a whole street at an intersection, and every manner of death possible to human beings displayed by corpses. When I involuntarily looked away, my brother scolded me, 'Akira, look carefully now' . . .
>
> In some places the piles of corpses formed little mountains. On top of one of these mountains sat a blackened body in the lotus position of Zen meditation. This corpse looked exactly like a Buddhist statue. My brother and I stared at it for a long time, standing stock still. Then my brother, as if talking to himself, softly said, 'Magnificent, isn't it?' I felt the same way . . .
>
> The night we returned from the horrifying excursion I was fully prepared to be unable to sleep, or to have terrible nightmares if I did. But no sooner had I laid my head on the pillow than it was morning. I had slept like a log, and I couldn't remember anything frightening from my dreams. This seemed so strange to me that I asked my brother how it could have come about. 'If you shut your eyes to a frightening sight, you end up being frightened. If you look at everything straight on, there is nothing to be afraid of.' Looking back on the excursion now, I realize that it must have horrifying for my brother too. It had been an expedition to conquer fear.[4]

Before the Great Kanto earthquake, the most serious earthquakes to ravage Tokyo (then known as Edo) were those of 1703 and 1855, the second of which – the Ansei earthquake – provoked the popular outpouring of catfish prints (*namazu-e*) discussed in Chapter 1. By the end of the 19th century, after the arrival of seismology in Japan, the leading question in Japan was inevitably: when would Tokyo again suffer another great earthquake? In 1905, this issue created a 'rift' between Fusakichi Omori, the former protégé of Milne who was now professor of seismology at the Tokyo Imperial University, and Imamura, who was assistant professor. The quarrel, in the words of historian Gregory Clancey, 'remains legendary among contemporary Japanese seismologists'.[5]

Although junior to Omori in position, Imamura was only two years younger than him, and the two men quickly became rivals, with rival theories of the origin of great earthquakes.

In Omori's opinion, the risk to Tokyo was reduced, not enhanced, by the frequent seismic activity on geological faults beneath the capital, since he thought that these smaller earthquakes must release the potentially dangerous build-up of seismic stress. Instead, Omori's suspicions focused on areas where there had been long gaps in seismic activity, such as the Nobi Plain, which had been comparatively quiet for hundreds of years before the Mino–Owari earthquake in 1891. Imamura, by contrast, focused on Sagami Bay, southwest of Tokyo, where there was a disturbing absence of seismic records because the geological faulting was located underwater. In an article for a popular journal in 1905, Imamura went so far as to predict a great earthquake in Tokyo within fifty years and advised that the city should be ready for a worst-case scenario. Moreover, he argued that given Tokyo's largely wooden construction, a great earthquake would cause a fire with an excess of 100,000 casualties.

Imamura's prediction, despite its prescience, had no scientific data to support it. Omori therefore publicly denounced Imamura's idea in an article in the same journal, entitled 'Rumours of Tokyo and a Great Earthquake', comparing the apocalyptic prediction with a popular 'fire horse' legend that conflagrations would occur during years in which there was an alignment of the astrological symbols for 'fire' and 'horse'. 'The theory that a large earthquake will take place in Tokyo in the near future is academically baseless and trivial', Omori asserted.[6] He also berated Imamura in front of his students on campus. But although the government's department of education ordered Imamura to recant, he refused to back down, although he was now shunned by some of his scientific colleagues.

In 1915, the two rival seismologists once again clashed publicly over Imamura's prediction. This time, Imamura had to leave his position at the Tokyo Imperial University for a while; when he returned to his home village, even his father censured him.

Meanwhile, Omori's reputation increased. His design of a seismograph, known as the Bosch-Omori seismograph (after its German manufacturer Bosch), was widely used throughout the world in the early 20th century. Not only was Omori hailed as a great seismologist in San Francisco, he also made predictions of earthquakes around the Pacific rim, in southern Italy and in China, which appeared to be borne out by seismic events in the Aleutian Islands and in Valparaiso in 1906, in Messina in 1908 and Avezzano in 1915 and in Haiyuan in 1920 – although it should be emphasized that Omori predicted the locations, not the dates, of these earthquakes. In his native Japan, however, Omori's theory of seismic gaps and seismic release would now turn out to be a terrible failure.

In late 1921, the strongest earthquake in Tokyo for twenty-eight years damaged a conduit and almost cut off the city from its water supply. An even stronger one in mid-1922 damaged buildings, cut the phone service and stopped the railways. Then, in early 1923, there was a third shock, less severe than the previous two.

Pursuing his theory, Omori maintained that these shocks, especially the third one, had demonstrated the relief of seismic stress in the faults beneath Tokyo. The capital could now relax, he announced. In a scientific paper published in 1922, he speculated that 'probably the semi-destructive earthquake on 26 April 1922, has finished the activity epoch succeeding the period of the seismic rest during the last half a dozen years.'[7] The weaker shock in early 1923 reinforced his conviction, and he wrote: 'Tokyo may be assumed to be free in future from the visitation of a violent earthquake like that of 1855, as the latter shock originated right under the city itself, and as destructive earthquakes do not repeat from one and the same origin, at least not in the course of 1,000 or 1,500 years.'[8] By the time that this paper appeared in print, in 1924, the Great Kanto earthquake had destroyed most of Tokyo, and Omori himself was no more.

On 1 September 1923, the day of the earthquake, Omori was far away from Tokyo in Sydney, Australia, at the Second Pan-Pacific Science Conference. He was being shown a seismograph by an Australian seismologist, when it suddenly sprang to life and recorded the distant destruction of his home city. Initial news reports mentioned tens of thousands of deaths, but Omori told Australian reporters that these figures were probably exaggerations – as had happened in the initial reports of 100,000 deaths in the Ansei earthquake of 1855 – before boarding the first available ship to Tokyo. The Australian journalists were not convinced. A reporter from the Melbourne newspaper, *The Age*, remarked that despite Japanese seismologists such as Omori having tried their best at earthquake prediction, 'the present horror is a sad commentary on their labours'.[9]

Omori's colleague and rival, Imamura, was at his desk in the Seismological Institute of the Tokyo Imperial University when the shaking began. He later described the experience:

> At first, the movement was rather slow and feeble so that I did not take it to be the forerunner of so big a shock. As usual, I began to estimate the duration of the preliminary tremors. Soon the vibration became large and after 3 or 4 seconds from the commencement I felt the shock very strongly indeed. 7 or 8 seconds passed and the house was shaking to an extraordinary extent, but I considered these movements not yet to be the principal portion . . . The motion went on to increase in intensity very quickly, and after 4 or 5 seconds I felt it to have reached its strongest. During this epoch, the tiles were showering down from the roof making a loud noise, and I wondered whether the building could stand or not.[10]

Soon after this, the institute's seismographs were physically overturned by the shaking, though not before registering the initial vibrations reported by Imamura, and the walls of the university building began to collapse.

Imamura and colleagues became fire-fighters, without water or help from outside, as they desperately attempted to save the seismological records of half a century, going back to the days of their British predecessors Milne, Ewing and Gray, from incineration.

Imamura survived the earthquake unscathed. Whether by luck or by good judgment, he had the grim satisfaction of seeing his 1905 prediction come true – not only in respect of the timing of the earthquake (well within his fifty-year window) and the scale of the disaster (well over 100,000 deaths from fire) but also in the location of the earthquake, which had its epicentre under Sagami Bay. He also, thanks to Omori's chance absence in Australia, became a crucial scientific adviser to the Japanese government and a leading spokesman to the world's press.

Greeting his senior colleague at the devastated dockside on Omori's return from Australia, Imamura reportedly received an apology. But by then, Omori was a sick man, suffering from a brain tumour. When Imamura visited him in hospital in late October, he found him wracked by pain and tortured by guilt. Imamura tried to comfort his colleague with a joint toast in glasses of sake to their university department of seismology. Omori died a week or two later, at the age of only fifty-five. 'It was the end of a life and career hardly capable of continuing as before', concludes Clancey's compelling account of the Omori–Imamura dispute in *Earthquake Nation*.[11]

It was also a foretaste of the bitter, still unresolved, controversies that would surround earthquake prediction in Japan, the United States and every other 'earthquake nation' throughout the rest of the 20th century and up to the present day, as we shall see.

The shock waves radiating from beneath the ocean floor of Sagami Bay struck Yokohama first, at exactly 11.58 a.m. They hit Tokyo, a little further from the epicentre towards the northeast, forty-four seconds after Yokohama. As in Lisbon in 1755 and in Edo in 1855, the aftershocks continued, through the day and night, over several days. At the Tokyo Imperial University, Imamura detected more than 171 aftershocks between 11.58 a.m. and 6 p.m., and a further fifty-one before midnight on 1 September. Today's estimate of the magnitude of the main shock is 7.9–8.1: slightly more powerful than in California in 1906.

Noboru Oshima, a reporter for the Jiji news agency in Yokohama, was lucky. He was in the office at 11.58 a.m. when he heard a sound 'like a distant detonation'. The next moment, he and his chair were tossed well above the floor and then landed face down on the ground. The building was swaying violently and groaning sinisterly, but somehow Oshima got to the head of the stairs – where he was promptly thrown down the steps, rolling head first. 'I was up in an instant and leapt out of the building, only to fall three feet into a fissure which the shaking had made in the concrete sidewalk,' he reported. When he managed to get to his feet and look around, he saw that every house, so far as the eye could see through the brown-black sky swirling with clouds of dust and smoke, had collapsed. 'The foreign houses on the bluff were already afire, sending up black smoke columns. The people in the street were all tottering.'[12]

In Tokyo, Randall Gould, a reporter with the *Japan Times*, was similarly fortunate. He was about to eat a sandwich lunch when the shaking started. 'Most of the office came down,' he said, 'but not all of it.'[13] Grabbing his typewriter, he scrambled into the street, where he saw that few other buildings were left standing. The city's trams had stopped in their tracks. At the Akasaka Palace, where the prince regent, Hirohito, was eating his lunch, there was little damage, because the palace was bolted onto massive concrete foundations. The prince ran into the extensive gardens, where he saw column after column of black smoke rising from the fires already burning in the city. They would continue to burn for more than forty hours.

The people of Tokyo immediately headed for the open spaces in the city. They had centuries of experience of great fires in the capital, which were so frequent that they had become known as *Edo no hana*, the 'flowers of Edo'. In 1923, the flames pushed some of the crowds back towards the imperial palace. Caught between certain death by fire and an armed confrontation with the police guarding the palace, the people forced their way into the safety of the palace's outer gardens and encamped there for several days.

Other people fled east towards the Sumida River, hoping to reach the supposed safety of the opposite bank. But the Eitai Bridge had been almost entirely wrecked by the earthquake, leaving only a single iron beam spanning the river, high above the water. There was no choice for the refugees but to start crossing the beam in single file.

What happened next, as extracted from reports and eyewitness accounts published in the *Japan Times* during the following weeks, was a panic in hell:

> A wall of flame was now racing eastwards, fanned by the strong winds. As it began to outrun those at the rear, the crowds heading for the Eitai Bridge panicked and surged forwards, crushing and suffocating those in front and sending 50 or 60 headlong into the water.
>
> Refugees crawling, ant-like, along the iron girder looked down on a horrific scene far below. Hundreds of people were in the water, some clinging to flotsam, others drowning, many already dead. Boats worked as hard as they could to ferry survivors across, but their task was hopeless in the face of the panicking thousands who lined the bank.
>
> One survivor, a woman, later told how she was pushed into the water but managed to cling to a rope moored to the bank. As the afternoon wore on, the heat from the flames grew more intense, and she found herself having to duck under the water every time a searing blast of wind hit her face. The river grew warm, and then frighteningly hot.[14]

At another point on the Sumida River, in the area of Nihonbashi, the flames leapt over the water from one bank to the other – a distance of 220 metres (240 yards). Those who survived their submersion in the filthy river finally pulled themselves out, singed and blackened, on the morning of 3 September, after nearly two days without food. Elsewhere, in the canals of the city, many people were boiled alive.

However, the most dreadful disaster by far occurred in the crowded, working-class, manufacturing area of Honjo. In 1920, its population was officially 256,269 people; in 1925, after the earthquake and fire, the number had fallen to 207,074. Some 40,000 inhabitants of Honjo perished on 1 September 1923 in one concentrated conflagration, which is still memorialized in Tokyo.

On the advice of the Honjo ward police chief, locals had taken refuge – along with the highly flammable furniture and baggage from their homes – in one of the very few open spaces in Honjo: a vacant site that until 1922 had been a depot for military uniforms but at the time of the earthquake was being transformed into a park by the municipal administration. Its area was 6 hectares (15 acres), that is, about six times the size of London's Trafalgar Square.

However, the space was too small to protect anyone from the raging flames that now approached it from several directions and effectively surrounded the screaming refugees. Inside the park, the flames became vortices up to 200 metres (650 feet) in height, spinning anticlockwise with a velocity of some 240 kilometres (150 miles) per hour (according to later investigations by Imamura). In *Tokyo: City of Stories*, journalist and geographer Paul Waley pictures the awesome scene that transpired after sunset on 1 September:

> Strong gusts of wind buffeted against the wall of flames creating a series of little whirlwinds that sucked people into the air and then dropped them down again as balls of fire. The whole park turned into a blaze of infernal proportions, hot enough to buckle steel and melt metal. Nearly everyone who had fled there was burnt to death, and afterwards the devastation was so complete that it was impossible to tell how many people had died.[15]

Of the handful who did get out, one was the eldest son of the managing director of the Tokai Bank. His father, Genjiro Yoshida, had taken his entire family to shelter in the former depot. They all died in the firestorm, except for Genjiro Junior, who had a miraculous escape: he was picked up by one of the fiery whirlwinds, whisked away and then dropped in a ditch, where he somehow survived the flames. On the night after the holocaust, the Honjo ward police chief committed suicide in shame and remorse.

Corpses littered downtown Tokyo, as recalled by Kurosawa decades later. According to the report of a correspondent from an Osaka newspaper, who flew over the ruins of Tokyo in an open-cockpit army reconnaissance plane, even at a height of 1,000 metres (3,300 feet) the odour of death overpowered the smell of engine exhaust, causing both the pilot and his passenger to retch.

On the ground, the well-known modernist writer, Ryunosuke Akutagawa (author of two stories on which Kurosawa based his film *Rashomon*), observed the aftermath and wrote about it in an autobiographical sketch, 'The Life of a Stupid Man'. Here he describes himself (in the third person) on a visit to the red-light area of Tokyo, Yoshiwara, where nearly 500 men and women – mostly courtesans – had died, either from incineration by glowing ash from surrounding buildings, which set fire to their hair, lacquered with oil and wax, or from suffocation in a pond in which they had thrown themselves in order to evade the flames:

> The odour was something close to overripe apricots. Catching a hint of it as he walked through the charred ruins, he found himself thinking such thoughts as these: *The smell of corpses rotting in the sun is not as bad as I would have expected.* When he stood before a pond where bodies were piled upon bodies, however, he discovered that the old Chinese expression, 'burning the nose', was no mere sensory exaggeration of grief and horror. What especially moved him was the corpse of a child of 12 or 13. He felt something like envy as he looked at it, recalling such expressions as 'Those whom the gods love die young.' Both his sister and his half-brother had lost their houses to fire . . . *Too bad we didn't all die.*
>
> Standing in the charred ruins he could hardly keep from feeling this way.[16]

Akutagawa's sketch was published posthumously after his suicide in 1927. His fellow writer, Yasunari Kawabata, who accompanied Akutagawa on his walk through the ruins of Tokyo, was convinced that seeing the ugly horror of the corpses in the pond had made Akutagawa determined to leave behind 'a handsome corpse'.[17]

Kawabata was a subtle writer, who eventually became Japan's first Nobel laureate in literature. In the 1920s, he wrote a fictional response to the earthquake in the form of a short story, one of those known as his 'palm-of-the-hand' stories, which has been translated into English as 'The Money Road'. According to the opening, it takes place on 1 September 1924. Its chief setting is the former military clothing depot in Honjo, which had by now become the site of a charnel house, marking the site of the authorities' mass cremation of the victims of the fire.

'On the anniversary of the earthquake an imperial messenger appeared at the ruins of the clothing depot', writes Kawabata.

> The prime minister, the interior minister, and the mayor all read memorial addresses at the ceremony. Foreign ambassadors sent memorial wreaths.
>
> At 11:58 all traffic stopped, and the people of the city observed a moment of silence.
>
> Steamships that had gathered from Yokohama made the trip back and forth between Kokokashiko and the bank near the clothing depot on the Sumida River. The automobile companies vied to be first to make an official appearance in front of the clothing depot. Each religious organization, Red Cross hospital, and Christian girls school sent a relief committee to the ceremony.
>
> A postcard dealer rounded up some vagrants and dispatched a squad to secretly sell photographs of bodies mangled in the earthquake. A movie studio technician walked around with a tall tripod. Money-changers stood in a row to change the visitors' silver coins for lesser copper coins to be tossed into the offering box.

Ken, a shrewd vagrant and beggar in worn-out army boots, dragging along an unnamed woman friend in sandals – a penurious older beggar, whose entire family was burnt to death in the clothing depot – join the crowd of mourners. The woman plans to offer a red comb in memory of her dead daughter. Taking off one of his boots, Ken gives it to her to wear, without explaining why, and they edge forward, each walking with one bare foot and one booted foot.

'Just as a brilliant forest of floral wreaths and funeral greenery came into view, their feet suddenly felt cold. It was coins.' The crowd of people, unable to reach the offering box, had started to throw money from wherever they stood; and the coins were falling on everyone's heads like hail. Both Ken and the woman immediately start picking up coins between their bare toes and dropping them into the two boots. 'The closer they advanced to the charnel house on the cold money road, the deeper the layer of coins grew. People were walking an inch off the ground.' In the end, they excitedly hobble away in their coin-filled boots.

Only then, as they sit together on the deserted bank of the river, does the woman remember that she forgot to offer the red comb. Emptying her boot of coins, she places the comb inside it and flings both into the river. 'The red comb floated out of the sinking boot and silently drifted down the great river.'[18]

In today's Tokyo, there is a park on the site of the former clothing depot in Honjo. In the middle, sheltered by trees from the incessant traffic, stands a temple dedicated to the memory of the victims of the fire. In the gardens around the temple stand strange, twisted sculptures, rather than the usual Japanese rock garden. Close up, they can be recognized not as works of modern art but as erstwhile metal machinery, such as presses and engines, melted by the heat of the firestorms in 1923.

Little else survives of the 1923 earthquake and fire. The evidence was destroyed by the reconstruction of Tokyo and Yokohama in the 1920s, then obliterated by the massive incendiary bombing of Tokyo during the Second World War in 1945. Indeed, United States military planners deliberately based their aerial attack on the destruction caused by the post-earthquake fire in 1923. The caption to an American wartime propaganda image, showing an aerial photograph of earthquake-devastated Tokyo, is chilling: 'For this is the house that Jap built. And this is its logical fate. What nature did, in levelling Tokyo to the ground with an earthquake, we must now do again with American guns and bombs.'[19]

Also obliterated was a grand plan for reconstruction conceived by the home minister of the imperial government, Goto Shinpei, a former mayor of Tokyo, in the immediate wake of the 1923 disaster. He saw the destruction as a blessing in disguise, a chance to clear away Tokyo's burgeoning slums and remodel the city on a European-style grid as a capital worthy of a great power. But other Japanese politicians, including the finance minister, were in no doubt that such an all-encompassing plan would cost far more than the nation could afford, and therefore axed Shinpei's proposed budget. A third political group, which represented impoverished rural regions far from Tokyo, resented the idea of massive spending on the capital and forced a further reduction in the budget. As for the residents of the burnt-out districts, they were largely in favour of rebuilding exactly what they had lost, and began doing so within days of the disaster.

In late 1923, the home and finance ministers clashed spectacularly in cabinet, and the home minister lost his case. He died a broken man in 1929. But Shinpei's plan was not utterly forgotten. In 1983, on the sixtieth anniversary of the earthquake, the Emperor Hirohito remarked that the plan, if implemented in 1923, 'might have reduced considerably the wartime fires of 1945 in Tokyo. I now think that it was very unfortunate that Goto's plan was not put into action.'[20]

In practice, two-thirds of the reconstruction budget spent between 1923 and 1930 by the city of Tokyo and the national government – approximately 744 million yen – went not on creating a brand-new Tokyo but on improving the roads, canals and bridges of the old Tokyo, along with a process of 'land readjustment'. This entailed a street-by-street negotiation with residents, who had to sacrifice up to 10 per cent of their private land without government compensation in the interests of bettering their city – largely by eliminating narrow alleyways in favour of straight modern roads with pavements. Local feelings were often outraged in the process, sometimes necessitating religious intervention, such as when a sacred tree had to be cut down. A Reconstruction Bureau official noted, probably wryly: 'It was a wonderful time for Shinto priests to gain quick riches.'[21] By 1930, Tokyo was officially declared to be reconstructed – even more quickly than happened in San Francisco after the 1906 earthquake.

As for the long-term effects of the Great Kanto earthquake, opinions differ. Three years after the event, the official report noted that: 'The almost total destruction of Tokyo, the capital of the Empire, and the complete destruction of Yokohama, the foremost of our leading ports, inflicted upon the nation a cruel wound and one not easy to heal.'[22] Ten years later, the total cost was estimated by the Tokyo Municipal Office at about 5.5 billion yen or $1.65 billion (in 1933 dollars).

It is not straightforward to relate this cost to the nominal gross national product of Japan in the years after 1923. 'One estimate places Japan's gross national product in 1930 at 13.85 billion yen', notes historian Gennifer Weisenfeld in her study of the earthquake, *Imaging Disaster*. However, 'The yen-dollar exchange rate fluctuated wildly in the 1930s because of the unstable economy, and the GNP also fluctuated during the depression, so it is hard to get a sense of equivalents.'[23] Although it is certain that the earthquake did not cripple Japan's economy, it undoubtedly destabilized it. According to historian Edward Seidensticker in *Tokyo Rising: The City Since the Great Earthquake*, debts stemming from the earthquake played a direct role in a financial panic and bank run in 1927 that led to the resignation of the cabinet and the appointment of an army general as prime minister who advocated aggressive interventionism in China. However, Seidensticker questions the existence of a direct link between this financial and political crisis and the militarization of Japanese society in the 1930s: 'Whether or not the reaction of the thirties would have come had the depression not come, we will never know.'[24]

A complicating factor was growing distrust and tension between Japan and the US, which had been simmering ever since the initial Japanese–American encounters in the 1850s. In September 1923, the Japanese authorities in Yokohama at first rejected aid offered by US ships in the harbour by jamming their radio messages. Then, Japan accepted a major American military relief effort with such gratitude that, after it was concluded, the US admiral in charge informed navy officials in Pearl Harbor there was no possibility of war with Japan in his generation. The American naval attaché in Tokyo during the earthquake was more pessimistic, however, stressing that the relief effort had aroused Japanese envy as well as gratitude. In the US itself, long-standing anti-Japanese feeling was fed by allegations that the Japanese had failed to show enough appreciation of the American relief effort. In mid-1924, the US Congress enacted racist legislation, the so-called Japan Exclusion Act, which naturally fed militarism, fascism and xenophobia in Japan. After the Wall Street Crash of 1929, against a background of worldwide economic depression, there occurred the Japanese invasion of Manchuria in 1931 and Japan's military adventurism in the rest of China, the military takeover of the Japanese government after the assassination of a liberal prime minister in 1932, the signing of a pact with Nazi Germany in 1936, and finally Japan's entry into the Second World War in 1941.

While it is reasonable to postulate a causal connection between the massive disruption of the earthquake and the eventual declaration of total war by Japan in 1941, it is more difficult to substantiate it. The consensus among recent historians and political commentators is that the martial law imposed between 2 September and mid-November 1923 gave new authority to the police and the army, which some officers exploited for their own agendas, including the massacre of Korean immigrants. The earthquake 'paved the way for political domination by the Japanese military', writes Richard Samuels. 'The military had rallied a vulnerable nation under banners of leadership, social solidarity and change'.[25] It was a 'turning point',[26] agrees David Pilling, which 'helped crystallise a lurch towards totalitarianism.'[27] The declaration of martial law 'gave an already aggressive military new power and stature in Japanese society', confirms Joshua Hammer. 'Many of the officers who rose to positions of authority in the earthquake's aftermath would play prominent roles in the radical anti-democratic groups that formed in the late 1920s and early 1930s, the same groups that would lead the country to war.'[28]

Instead of political influences from the earthquake, Seidensticker – who is best known as a scholar and translator of Japanese literature – discerns subtle, if unproven, cultural influences. Before the earthquake, customers entering Tokyo department stores would automatically change their shoes for specially provided slippers; after the earthquake, they could enter in their ordinary footwear. Henceforth, Japanese department stores became more like their equivalents in New York and London. At the same time, after the earthquake, Japanese women – more of whom were now working – began eating out in department-store dining rooms; before the earthquake, such public eating was not considered to be good form for a woman. Lastly, the particularly Japanese passion for cartoon strips and comics with a panel narrative, nowadays familiar as *manga*, dates from the years immediately after the Great Kanto earthquake. 'Whether or not their origins can be blamed on the confusion that followed the earthquake, that is where they are', observes Seidensticker.[29] Given the undoubted origin of another distinctively Japanese graphic form, *namazu-e*, in the events of the 1855 earthquake, a similar kind of artistic and literary phenomenon arising from the 1923 earthquake seems plausible.

However one interprets the Great Kanto earthquake's effects on Japan, they make a striking contrast with those of a great earthquake on China half a century later. The Tangshan earthquake in 1976 would not only catalyse the end of a dictatorship and the rise of a great economic power, but also mark the end of China's violent Cultural Revolution.

CHAPTER

8 BIRTH PANG OF A NEW CHINA: TANGSHAN, 1976

Improvised classroom in Tangshan in autumn 1976. The students learn: 'Father is good, Mother is good, but Chairman Mao is best'.

By a cruel irony, not long before the People's Republic of China suffered the world's most lethal earthquake of the 20th century, the country also benefited from the world's sole earthquake prediction that enjoys at least a modicum of credibility among seismologists: that of the Haicheng earthquake in 1975. No large earthquake (and relatively few moderate ones) had occurred for more than a century in the Liaoning province of Manchuria – where the town of Haicheng is located near the northern end of the Bohai Sea – when, during early 1974, minor tremors began to increase. In the first five months of 1974, Chinese seismologists measured five times the normal number. They discovered, too, that much of the region had been uplifted and tilted to the northwest, and that the strength of the earth's magnetic field was increasing in the area. They also observed anomalous underground electrical current readings and well water levels. The State Bureau of Seismology in Beijing – founded just three years earlier out of an initiative by Premier Zhou Enlai, following a strong earthquake not very far from the capital city in 1966 – issued a forecast: Liaoning should expect a moderate to strong earthquake within two years. On 22 December 1974, there was another burst of tremors. The forecast became more focused: expect an earthquake of magnitude 5.5–6.0 somewhere in the region of Yingkou, a major industrial port, during the first six months of 1975.

All over the affected area, animals began to behave strangely. Snakes awoke from hibernation prematurely and lay frozen in the snow; rats appeared in groups so agitated that they did not fear human beings; small pigs chewed off their tails and ate them. In addition, wells began to bubble. A swarm of tremors – 500 were recorded in seventy-two hours – culminated in a magnitude-5.1 jolt at 7.51 a.m. on 4 February 1975. A number of moderate shocks then followed, but by the evening the seismic activity was dying down.

Nevertheless, local Communist Party officials – after being vigorously spurred on by the head of the local Earthquake Office in Yingkou County, a former army officer named Cao Xianqing – decided on an evacuation, during an emergency meeting convened at 8.15 a.m. on 4 February. At 2 p.m., 3 million people were ordered to leave their homes and spend the night outdoors in straw shelters and tents. Without any panic, the masses of southern Liaoning province obeyed. The outside temperature was already many degrees below freezing. In Haicheng County, there was less sense of urgency from local observers of the tremors, and the evacuation was not so extensive.

In the course of the day, Cao predicted that the main shock would occur before 8 p.m., and furthermore, that it would be of magnitude 7 at 7 p.m. and of magnitude 8 at 8 p.m. At 7.36 p.m., the earthquake struck – with a magnitude of 7.3. Sheets of light flashed through the sky, the earth heaved, 4.5-metre (15-foot) jets of water and sand shot into the air. Roads and bridges buckled, rural communes tumbled down. The majority of the buildings in Yingkou and neighbouring Haicheng, a town of 90,000 people, were wrecked. But because of the afternoon evacuation of people from their homes, and also because of the earthquake's evening timing – when large masonry structures such as schools, office buildings and factories were empty – rather than a death toll of tens of thousands, there were relatively few victims: some 2,000 fatalities from the earthquake, fires and hypothermia.

According to the party line emanating from Beijing in 1975, 'The Haicheng earthquake was successfully predicted, saving untold thousands of lives.'[1] A careful later investigation, published in 2006 in the *Bulletin of the Seismological Association of America* by Kelin Wang, Qi-Fu Chen, Shi-hong Sun and Andong Wang, was more equivocal. It accepts that the energetic actions of Cao Xianqing in Yingkou were crucial in saving lives. However, it shows that he acted more from some gut instinct than from any scientific theory in predicting the precise details of the earthquake's timing and magnitude. 'This patently absurd aspect of the forecast was apparently based on some sort of extrapolation of foreshock activity', writes seismologist Susan Hough in her study of earthquake prediction, *Predicting the Unpredictable*.[2] Cao himself told interviewers that he had based his forecast on a Chinese book, *Serendipitous Historical Records of Yingchuan*, which stated that heavy autumn rains would certainly be followed by a winter earthquake. Observing the heavy autumn rains of 1974 and the escalating seismic activity in this period, he predicted an earthquake before what he thought was the official end of the Chinese winter at 8 p.m. on 4 February. But since he got this time wrong – the official end was actually at 6.59 p.m. – by his own reckoning the earthquake was half an hour late!

In stark contrast, a year and a half later, both nature and man conspired against the sprawling industrial city of Tangshan, 150 kilometres (90 miles) roughly to the east of Beijing, and its surrounding area. Tangshan was not notable for seismicity, either in modern times or in the historical record, although there had been strong earthquakes in the general region of Tangshan, around and beneath the Bohai Sea, in 1966, 1967, 1969 and 1975 (in Haicheng). Tangshan's magnitude-7.8 earthquake, more powerful than that in Haicheng, lasted for just twenty-three seconds between 3.42 and 3.43 a.m. on 28 July 1976, unannounced by foreshocks. The few precursors that had been observed, such as strange animal and insect behaviour, unexplained lights in the night sky and disturbances in well water levels, were either insufficient for any meaningful prediction or had been ignored by the authorities. For in 1976, the State Bureau of Seismology was in political turmoil, along with the rest of China. Premier Zhou Enlai had died in January; his chosen successor, Vice-Premier Deng Xiaoping – who had delivered the eulogy at Zhou Enlai's funeral – had been officially disgraced in April at the behest of Chairman Mao Zedong, for allegedly orchestrating public protests in memory of Zhou; the Maoist 'Gang of Four', led by Mao's wife, Jiang Qing, was aggressively in the ascendant; and Mao himself was lying on his deathbed in Beijing. On 12 July, a seismology conference to consider current earthquake risk in the Beijing–Tianjin–Tanghsan–Bohai Sea–Zhangjiakou region was held in Tangshan. Yet on the very same day, in Beijing, the group of party leaders in charge of the seismology bureau – who were not scientifically trained – held a political meeting, which severely attacked their own group leader.

The febrile atmosphere of this troubled period was distilled in a pioneering book published in 1986 in Chinese and later translated as *The Great China Earthquake*. Its author, Qian Gang, a cadre in the People's Liberation Army (PLA) and staff reporter for its newspaper, personally assisted in the rescue of Tangshan. Its graphic human details transcend its inevitable ideology. Even when dealing with politics, the book, though certainly not exempt from party propaganda, is not afraid to criticize some excesses. The substantial section on seismology was based on the already yellowing archives of the seismology bureau, recorded in 'language filled with the jargon of the Cultural Revolution', notes Qian.[3] Thus the 12 July 1976 meeting in Beijing produced an urgent report sent to the Academy of Sciences censuring the head of the bureau for 'keeping close step with Liu Shaoqi and Deng Xiaoping'. Hu Keshi, the report noted,

> has persistently pursued a revisionist line; in this rightist trend to reverse verdicts, he is clearly acting as a capitalist roader. Hence it is no longer suitable for him to lead the bureau's party leaders. He should be removed from his position as head of the leading group, write a self-criticism and subject himself to mass criticism.

As Qian frankly comments, 'In the turbulence of the "Gang of Four's" attempt to seize power, the State Bureau of Seismology was no refuge. Its office block, along with countless others in China, was choked with the stifling air of politics.'[4]

When, on 27 July, the deputy head of the bureau, a former military man, told the assembled seismologists, 'Well, you've talked long enough, what's your view? Which way is the Prediction Office leaning? Have you been able to identify a pattern?' he received no useful answer.[5] The great earthquake flattened Tangshan the following night, wholly unpredicted by the central government's seismology bureau.

Tangshan, unlike Haicheng, was a vital centre of coal mining, steel making, ceramics manufacturing and agricultural marketing, with a population of over a million people: a thousandth of the whole of China, but responsible for a hundredth of the country's economic output, so it was often said. The city was the home of China's first modern coal pit, its first standard-gauge railway (built to haul coal), its first steam locomotive and its first cement works – all of them originally constructed by foreigners in the later 19th century.

But there was to be no foreign help after the earthquake. The Chinese government flatly rejected immediate offers of aid from 'foreign devils' – that is, Japan, the United Nations, the European Economic Community and the

United States – and drew a cloak of secrecy over the disaster; Tangshan was closed to foreigners for some two years, including a delegation of American earthquake engineers officially invited to China in mid-1978. Nor was any official death toll announced in the months following the earthquake. Estimates of deaths in Tangshan and its surrounding area in foreign newspapers ranged between 650,000 and 800,000, based on the known population of the city and its apparently almost total destruction. In 1977, the Hong Kong-based *South China Morning Post* quoted a 'top secret' Chinese document with a figure of 655,237 persons killed and another 779,000 injured, 79,000 of them seriously. The official death toll was eventually given as 242,000, but this covered only the residents of Tangshan, excluding migrant workers, visitors and the surrounding area. The toll for every affected person is widely thought to be around the higher figure, even as high as 750,000. 'To this day, the true death toll remains a haunting unknown', notes a recent study by Beatrice Chen, a curator at the New York-based Museum of Chinese in America.[6]

Chen cites an incident that helps to convey the scarcely believable scale of the devastation. Just after eight in the morning of 28 July, some four and a half hours after the earthquake, a fighter jet landed at the PLA base 9 kilometres (5.5 miles) from Tangshan. Two army officers ran towards the plane. One of them, named Lee, asked the pilot: 'What is the flight's mission?' He replied: 'We are looking for the epicentre of the earthquake.' Immediately, without waiting to check the identity or the credentials of the other passengers on the jet, Lee asked the pilot to fly over Tangshan and confirm the army's suspicion that the epicentre lay under the city. The plane took off towards Tangshan. Lee radioed: 'Can you see Tangshan yet?' Through the loudspeaker the pilot's voice shakily responded: 'Yes, where it used to be!'[7]

Strange to tell, the safest place in Tangshan during the earthquake was underground. About 10,000 miners were at work when it struck. They were saved partly by the depth of the coal mines and partly by the high standard of the mines' construction, thanks to the expertise of their original western engineers working for the British and Belgian investors who started the mines in the 1870s. The Kailuan mining complex operating in Tangshan in 1976 was designed by Herbert Hoover – the future US president – when he was chief engineer for the Chinese Bureau of Mines around 1900, and nationalized when the People's Republic was founded in 1949. By 1976, it was something of a showpiece. Miraculously, only seventeen miners died in the complex during the quake, although five of them were not rescued until 11 August, having somehow survived underground for fifteen days without food. The miners' families above ground were not so lucky; one miner lost all of his eight children.

In the absence of working telephones, four staff from the mines at or near the surface were the first to bring the news of the catastrophe in Tangshan to the authorities in Beijing – a heroic episode described in dramatic detail by Qian Gang under the title, 'The Red Ambulance'. On 28 July, shortly after 4 a.m., he begins:

> A red ambulance came screaming out of the Tangshan mine at Kailuan. It made its way over piles of rubble and then, in the grey fog of dust that stretched as far as the eye could see, it gathered speed and, jolting and swaying, careered recklessly down the collapsed, uneven Xinhua Road, heading west. It was the first vehicle to reappear on the streets of Tangshan after the earthquake.[8]

The Kailuan miners' original idea had been to report the disaster to the local party headquarters in Tangshan. But they had quickly found this building to be in ruins, along with the army division headquarters, seven large hotels on the Xinhua Road, and apparently almost every building in the city, including all the hospitals but the one at the airport. A line of trees had been offset by up to 1.5 metres (5 feet). So, hardening their hearts against cries for help from a handful of wounded survivors, pursued by broken bricks thrown at their ambulance, the four miners headed out of the city as fast as the roads would allow, hoping to find a working telephone in one of the places between Tangshan and Beijing. However, it soon became obvious that obtaining a working phone line en route to Beijing would take them longer than actually driving there. And so, less than four hours after leaving Tangshan, the racing red ambulance turned up on the doorstep of Zhongnanhai, the government complex in Beijing, located in the series of palace complexes built by the Chinese emperors attached to the Forbidden City.

Its arrival was challenged by a policeman at the Xinhuamen Gate. One of the miners, Li Yulin, deputy director of the Tangshan Mine Union, jumped out of the ambulance wearing only his shorts, and announced: 'We're from Tangshan, we've come to declare an emergency to the State Council.'[9] The ambulance was immediately directed to the State Council reception centre. Here Li threw on an old jacket and was about to enter the centre when he saw that his hands were covered with blood: the blood of a dead neighbour in Tangshan whose child he had helped to dig out before departure. Squatting down by the side of the road, he washed his hands with water from a gutter, rubbed his face, and promptly went inside. The time was 8.06 a.m.

Shortly after, two airforce men, who had just flown in from Tangshan, arrived. They, Li Yulin and his mining colleagues were admitted to the council chamber in the Ziguang Pavilion. They saw an emergency meeting of vice-premiers in progress around a table on which was spread a large map, at which the party leaders were pointing with red pens. The atmosphere was very tense. Beijing itself had been badly shaken by the quake. Outside the city, the 500-year-old tombs of the Ming emperors were damaged: roofs shook and shrines collapsed, along with the guardian statues at their entrances. Within the city, most residents moved into their courtyards or into the streets, afraid to return to their houses. Deng Xiaoping and his family moved into a hastily erected tent next to their house. Indeed, Mao Zedong's hospital bed, after being vigorously shaken, had been wheeled through the pouring rain, surrounded by his medical team, into the recently erected Building 202 of Zhongnanhai, which was regarded as earthquake-proof.

Li Yulin recalled what happened next in an interview with Qian Gang a decade later:

> When they saw us coming in they all stood up. I said, 'Vice-premiers, there's nothing left standing in Tangshan!' Li Xiannian, Chen Yonggui and Ji Dengkui came up and embraced me. Somebody, I can't remember who, said, 'Don't get upset, just sit down, drink this and take your time telling us . . .'
>
> They were all asking, 'How is it there?' And when I began talking I started to cry.

> 'Vice-premiers, there are a million people in Tangshan; at the very least eight hundred thousand of them are still buried!'
>
> Some of the people sitting there began to weep.
>
> Li Xiannian asked me, 'How about the mines? How many people are there down there?'
>
> I told him, 'Ten thousand.'
>
> 'Ten thousand! That's terrible!' Then he asked me, 'Does Tangshan have mainly high-rise flats or one-storey buildings?'
>
> I told him, 'In the north of the city it's mainly high-rise, and in the south one-storey, so about half and half.'
>
> 'We have to rescue those people as quickly as possible!'
>
> Chen Xilian handed me a piece of paper and asked me to sketch out a rough map of Tangshan. Wu De came up and asked, 'The British set up a high-rise block for the Kailuan administration, whereabouts is that?' I pointed to the map and said, 'It's here. It's already collapsed.'
>
> Wu De gave a sigh. He knew Tangshan, had worked there as party secretary for the municipal party committee, and knew well the building he was referring to, the one the British had built – it could withstand any pressure, had walls a metre thick. 'Tangshan is gone, just gone . . .'[10]

Of the 11 square kilometres (4.2 square miles) of built space in Tangshan, 10.5 square kilometres (4 square miles) collapsed; less than 3 per cent of buildings in the centre of the city survived. The only bright spot was the absence of fire – presumably because of the timing of the earthquake in the early hours of the morning and the total power cut it caused.

How would the regime respond to such a stunning catastrophe? The distraught conversation on 28 July between the miners and the vice-premiers provides a significant clue. First and foremost, in the eyes of both the miners and the party hierarchy, Tangshan's mines and steel factories had to be restored to production. 'Tangshan mattered because of the mines and factories, not because of the people', observes journalist James Palmer in his insightful and moving study based on interviews with the earthquake survivors more than three decades later, *The Death of Mao: The Tangshan Earthquake and the Birth of a New China*.[11]

This task got underway within two or three days of the earthquake. Engineers from Tangshan and from outside worked twenty hours a day. On 9 August, the first of the eight major mines reopened. At noon on 11 August, engineers stumbled upon the five survivors from mine No. 5, and told them in blunt amazement: 'We figured you'd died a long way back'.[12] Later, the fortunate five were told that a young miner had heard their voices underground on 9 August but run away, thinking they were ghosts. By 16 August, six more mines were operating again, albeit at much less than full output. The achievement still stirred pride in some miners interviewed by Palmer: 'They couldn't do that nowadays!'[13]

The rest of Tangshan had to fend for itself, however, at least in the days immediately after the earthquake. Survivors, using only their bare hands, dug out the trapped; food, water, clothing and blankets were shared; cooperation and neighbourliness prevailed over competition and confrontation. 'On the morning of 28 July, Tangshan was the most truly communist place in the world', writes Palmer without irony. 'There were many examples of selfishness, greed and cowardice to be found that day, but they pale beside the extent of generosity, selflessness and courage.'[14] Fortunately, the city had no immigrant community who might have provided a focus for native fears and violent retribution, unlike the Koreans targeted in Tokyo in 1923. (The only foreigners killed as a result of the earthquake were three Japanese engineers employed in the mines.)

Fear and violence, however, were pervasive aspects of Mao's Cultural Revolution, especially in 1976, 'a year when everything was steeped in politics, smeared with political colours', comments Qian Gang.[15] One of the revolution's favourite sayings was: 'Wrongly killing a hundred people is better than letting a single guilty one escape.'[16] A writer and teacher, Feng Jicai, whose house in Tianjin, a major city 100 kilometres (60 miles) southwest of Tangshan, was destroyed by the earthquake, had particular reason for anxiety. He had long been noting down stories of the Cultural Revolution from torture victims and storing them within cracks of his house for security, covered by posters of Mao and the revolution – emulating a practice of Confucian scholars hiding their works from the persecutions of the First Emperor. Feng now faced the alarming possibility that these incriminating papers would be discovered by clean-up crews. Fortunately he had time, in the days after the quake, to rummage carefully through the ruins of his house in search of these writings. 'In the end,' he noted, 'I collected a whole bagful of small pieces of paper.'[17]

And yet, argues Chen, the fear instilled by Mao's policies must be counted as a useful factor in Tangshan's return to life. 'What made possible the reign of fear during the Cultural Revolution also allowed Tangshan's recovery process to develop efficiently . . . Without Mao's legacy of organizing and mobilizing the masses, disaster relief would likely have been chaotic and slow.'[18]

Such was the discipline instilled by the long years of party rule – despite the destruction of the local party machinery – that there seems to have been relatively little looting from the ruins. Not a single savings account in the ruined banks was lost, if one is to believe the later claims of the authorities. However, Qian Gang's book does honestly include a short section on 'Release of Criminal Energy'. For example, this canny thief:

> Someone saw an old woman weeping over a dead body, 'My son! My son!' Her weeping over, she plucked the watch off the dead man's hand and was gone. After a short while she reappeared before another corpse, and again there were tears, again came 'My son!' and again she plucked the watch. She moved thus from place to place, sobbing and taking a dozen or so times before she was caught.[19]

Presumably, the old woman was handed over to the Tangshan people's militia, and her stolen watches were among the 1,149 recorded as having been seized during the period of the earthquake in a report by the militia quoted by Qian.

On the other hand, it is difficult to know the true situation because Tangshanese were decidedly reluctant to address the topic of looting during Qian's research, 'as though these pages of history could, like the calendar pages of that week, be crumpled into a ball and thrown away.'[20] The manager of a market at the time, interviewed later by Qian, fell silent for a long while when asked about looting, then said: 'I can't quite call it to mind . . . It didn't really leave much of an impression on me.'[21] Historically, Qian notes, there is no record of looting by the people after major earthquakes in China, ranging from Shaanxi in 1556 to Haiyuan in 1920. Yet, there are frequent references to 'robbers rising up in swarms'. Were these robbers a form of 'abuse directed by the ruling classes against resistance shown by the common people', as was generally assumed in China after 1949?[22] Qian pragmatically leaves the question hanging in the air, with the implication that neither the common people nor the Communist authorities (including the people's militia) were as blameless of criminal activity as each of them seemed to suggest.

In support of the survivors, relief flights began landing at Tangshan airport, which had survived the worst of the destruction, on the day of the earthquake. The first PLA soldiers reached the city on the following day, 29 July, when telecommunications were also restored and the first small electric generator was got going, lighting a single street lamp, which became a beacon of hope in the pitch-black city. On 30 July, the State Council decided to transfer the seriously wounded to hospitals in eleven provinces and municipalities around China. An epidemic was avoided by spraying the city with pesticides from low-flying planes. But rescue work was severely handicapped by the PLA soldiers' dearth of appropriate materials and equipment. They lacked timber for shoring up collapsed buildings, and heavy rescue equipment such as cranes, electric saws, rock drills and electric-arc cutting machinery. 'They were an army without weapons,' writes Palmer, 'forced to resort to sawing through steel rods with nothing more than hacksaws, or using blasting caps from the mines to blow holes through fallen walls and hope they wouldn't bring the whole structure down.'[23] At dead of night, the soldiers would lie down on the ruins, hold their breath and listen for the slightest sound of a buried survivor. Not until ten days after the earthquake, on 7 August, did the required rescue equipment arrive: a stark commentary on the true importance accorded to the people of Tangshan by the party authorities in Beijing. Nevertheless, the city-dwellers fared infinitely better than those living in the countryside around the stricken city, where relief took months to reach the villages, if it reached them at all. Farmers, unlike industrial workers, were considered expendable by the regime; China had hundreds of millions of farmers.

On 9 August, a large crane operating on the wreck of a major hospital – while being keenly watched by China's news journalists, television and film cameramen – lifted up a 2-ton slab from the building and revealed a 46-year-old housewife, Lu Guilan, who had survived for thirteen days in the rubble, partly by biting her own tongue so that she could taste the moistness of her blood (as she later informed Qian Gang). Although she was too injured to move, when two soldiers brushed some clay off her eyelids, her first words were supposedly: 'Long live the PLA!'[24] Other survivors are said to have informed their rescuers that they owed their survival to their contemplation of the thoughts of Mao. One man is reported to have rescued the local party boss before going to search for his wife in the rubble.

Absent from these accounts – including those in *The Great China Earthquake* – are private experiences of trauma, grief and mourning. These emotions had no place in the party's ideology of resilience, stoicism and patriotism. The government announced no national day of mourning, nor did it really acknowledge the dead. Only after the reconstruction of Tangshan in the 1980s did the disaster acquire a more human face, which was certainly present in 1976 but was suppressed by both the people and the authorities. The most comprehensive photographer of the earthquake, Chang Qing, owed his survival to a fluke: he had swapped family apartments nine months earlier with a colleague and his family, who did not survive. When Palmer eventually asked Chang why he had not photographed the heaps of corpses, he replied: 'They were everywhere, but I couldn't bear to look at them.'[25] Then he began to cry softly. In a journal published in 2001, a survivor who lost a daughter in the earthquake, Chen Zhu-Hao, remarks that mourning the dead accompanied both the daytime search in the ruins for other survivors and corpses and the night-time existence among the ruins in tents made from wooden sticks and plastic sheets. His journal does not even mention the role of the Communist Party, remarks Beatrice Chen in giving her translation of this extract:

> It took only half a day to build our new house. We also found some pieces of timber and laid them on the ground so it would not be so wet. Since we didn't have many things, we did not need the entire plastic sheet so we left the unused

> portion on the ground. In about ten days, my neighbour Liu and his two daughters came back to Tangshan from the countryside. They only had one tent so Liu used the rest of our plastic sheet to build another tent for his daughters. Alone in the tent at night, I forgot the misery of recovering the bodies under the scorching sun but only to be occupied by other concerns:
>
> This is my home, I'm going to be living here, my new home, a new beginning . . .
>
> Will my eldest daughter be safe in the countryside?
>
> We were a family of five, but now there are only four of us left. Here, I'm all alone sleeping in the tent . . . another one lying in the dirt.[26]

At this desperate time for the people, the sole high-level party leader to visit Tangshan was Vice-Chairman and Premier Hua Guofeng, who was now the designated successor of Mao. Not surprisingly, the official press gave the visit wide publicity, with photographs of him meeting the earthquake victims. But in addition, the visit went down well with the people of Tangshan because of Hua's informality. Photographer Chang Qing, who covered it in detail, recalled more than three decades later to Palmer that Hua was 'easy to get along with. A little soft, not a hard man like many of the politicians. And he never insisted on the official tunes being played when he arrived; he just walked into the room and started talking to people.' At the end of the visit, genuinely eager crowds came to see Hua off to Beijing at the train station, photographed by Chang. 'When he came, it was the first time I felt really hopeful about the future of our city.'[27] Nonetheless, Chang was unwilling to give photographs of Hua's visit to Palmer, because they included the faces of too many officials who were purged after the ousting of Hua in 1978 by Deng Xiaoping. For the same reason, no doubt, there is not one single reference to Hua's visit in Qian Gang's 1986 book about the earthquake.

Hua's official report on the earthquake, delivered to Mao Zedong on 18 August, was the last official document read by him. Mao passed away in Beijing just after midnight on 9 September. When his death was finally announced at 4 p.m. that day, mourning began across China, but not with the same intensity as there had been for Zhou Enlai in January, and not much at all in devastated Tangshan. Instead, 'Like seismologists poring over charts of ground anomalies, the Chinese public looked for signs of the political earthquake they feared was coming'.[28]

In Tangshan, there was a persistent rumour that the State Bureau of Seismology had issued a correct prediction of the earthquake. This was said to have been concealed as a result of a power struggle among those around the dying Mao. The motive for suppression was thought to be either that elements in the government were set on provoking a natural disaster that would allow them to stage a political coup – echoing a traditional Chinese belief that natural disasters, when badly managed, have presaged the fall of dynasties – or simply that the government was indifferent to the fate of the people of Tangshan. Ironically, 'The government's own insistence on the Haicheng prediction as a miracle of scientific socialism, submissive nature bowing before Maoist ingenuity, came back to haunt it', writes Palmer.[29] If the science worked at Haicheng in 1975, then why not at Tangshan the following year? Seismologists in Tangshan were treated with venomous hostility as a result of their failure to predict the earthquake. The head of the seismology bureau's local prediction office had been badly injured in the disaster and spent three days lying in a pile of corpses at the airport, wrapped in a rain-soaked quilt. When he was at last taken for medical treatment, a crowd of injured angrily shouted: 'Don't treat him, Doctor! They've got nerve, still living! Good-for-nothings! Let them bleed to death! Why didn't the earthquake get them!' Although treatment was given, even the doctor asked the wounded seismologist: 'Comrade, why was there no prediction?'[30]

While there was little, if any, truth in the feasibility of accurate earthquake prediction by Chinese (or indeed any) scientists, there was undoubtedly a power struggle, between the government led by Hua Guofeng, the Gang of Four and the followers of Deng Xiaoping, who was still officially disgraced. Mao's death on 9 September intensified the struggle. But without the catalytic effect of the earthquake disaster, both this short-term political struggle, and the long-running national self-destruction of the Cultural Revolution, might well have dragged on for far longer than they actually did, stymieing the much-needed reform of Mao's failed political and economic policies.

Compared with Hua, the Gang of Four fatally misread the shock of the earthquake. Not only did none of them trouble to visit Tangshan, they also viewed the disaster in rigidly political terms. In public they said: 'Be alert to Deng Xiaoping's criminal attempt to exploit earthquake phobia to suppress revolution! Solemnly condemn the capitalist roaders who use the fear of an earthquake to sabotage the denunciation of Deng!'[31] In private they were alleged to have said: 'The earthquake in Tangshan affected only one million people, of whom only a few hundred thousand died. It's nothing compared to the criticism of Deng, which is a matter of eight hundred million people.'[32] Earthquake relief was regarded by them as a diversion from the true revolutionary path, even a cover for counter-revolutionary activities. But in 1976, unlike in 1967 at the launch of the Cultural Revolution, the country was disenchanted with revolution, distrustful of radicals and yearning for economic progress rather than ideological purity.

As events transpired, there was also a kind of political coup in the wake of the earthquake. On 6 October 1976, less than a month after the death of Mao, Hua Guofeng, in collaboration with other party leaders, had the Gang of Four unexpectedly arrested and in due course charged with treasonable activities. He also declared a formal end to the Cultural Revolution.

Moreover, there was truth in the rumour that the government was indifferent to Tangshan. Government support for rebuilding the city proved to be disastrously inadequate, despite media propaganda to the opposite effect, which assured the Chinese public that the citizens of Tangshan, thanks to the party, were now enjoying a higher standard of living than they had before the earthquake. 'After the initial fund-raising campaign, there were no channels for public donations', notes Palmer. 'Word of the real extent of the devastation spread by mouth, prompting many to send private gifts to relatives or friends in the destroyed city, or to use their connections to arrange for their families to be moved to other provinces.'[33] In the meantime, the government proceeded with the building of a vast mausoleum for Mao Zedong in Beijing's Tiananmen Square, with memorial calligraphy by Hua Guofeng. By May 1977, it was ready for the public.

Not until 1979–80, under the regime of Deng Xiaoping, did rebuilding of houses begin in Tangshan; as late as 1983, two-thirds of earthquake survivors remained refugees in their own city. Nevertheless, reconstruction was essentially complete by 1986, ten years after the earthquake, in time for Tangshan's coal mines to fuel factories making cheap toys and plastic goods for an export boom. However, the delay proved advantageous in some ways. Not only were the new buildings – mostly low-rise, without skyscrapers – designed to be earthquake-resistant, they were also designed by professionals, rather than by party officials. 'For the first time in modern China, the central government seemed willing to delegate its political authority and leave the planning of the city to experts', comments Chen. 'As a result, the plan for new Tangshan was created by experienced planners and academics more interested in building a city than in constructing a particular ideology.'[34] Neither Marxist theory nor market capitalism shaped today's Tangshan, in contrast to most other major Chinese cities.

The earthquake itself is remembered in Tangshan in the form of a rather brutalist monument from the 1980s; a 21st-century museum emphasizing both the heroism of individual survivors (including a model of the legendary red ambulance) and the superior wisdom of the Chinese Communist Party; and a memorial wall, finished in 2010, which is inscribed with the names of every known victim.

The monument and the museum are unsurprising – but the memorial wall, however familiar this idea may be in the West, is unique in China, according to Palmer, in that it recognizes 'individual loss, rather than collective sacrifice'.[35] This contrast perhaps encapsulates the chief historical legacy of the Tangshan earthquake. Immediately after it occurred, a credible political slogan was: 'An Earthquake is an Education in Communism'.[36] Yet in the longer term, the Chinese people disagreed with the slogan. The aftermath of the earthquake revealed that the people put their faith more in individual effort, expertise and reward than in Mao's ideology of collectivization. Deng Xiaoping, though almost as authoritarian as Mao, was astute enough to see that he could not buck this popular trend after he took over power from Hua Guofeng. Therefore, 'He allowed it to proceed, using economic growth instead to consolidate the party's grip on power, so badly eroded during the Cultural Revolution', writes a leading historian of the period, Frank Dikötter.[37] Thus the Tangshan earthquake, in hindsight, proved to be a birth pang in the painful creation of a wealthy and powerful new China.

CHAPTER

9 GRIEF AND GROWTH IN THE LAND OF GANDHI: GUJARAT, 2001

Statue of Mahatma Gandhi in Gandhidham, Gujarat, after the earthquake in 2001.

Authoritarianism was challenged in Communist China by the Tangshan earthquake in 1976. In democratic India, by contrast, authoritarianism was reinforced as a consequence of an earthquake in the state of Gujarat in 2001. In China, the earthquake catalysed the end of Maoist policies. In Gujarat, it led directly to the rise of a Hindu nationalist leader, Narendra Modi, who was elected prime minister of India in 2014. In both China and India, the earthquakes served to promote industrial growth.

The Indian earthquake struck on India's Republic Day, 26 January, in 2001, in the western part of Gujarat – the large, remote, coastal district of Kutch on the Arabian Sea – close to the tense border with Pakistan. Its epicentre lay only 20 kilometres (12 miles) from the district's headquarters: the historic town of Bhuj (which sometimes lends its name to the earthquake), founded in the mid-16th century. The shaking was felt throughout northwest India, including in the country's capital, Delhi, and in much of Pakistan, as well as in western Nepal and even in Bangladesh.

With a magnitude of 7.7, just shy of the magnitude in Tangshan, the Gujarat earthquake was, however, far less catastrophic in effect than the Chinese quake. Kutch was comparatively underpopulated and undeveloped; indeed, much of the district still consists of a vast and empty salt marsh, the Rann of Kutch, bordered in the west by the delta of the Indus River in the Pakistani province of Sindh and in the north by the Thar desert in India. Nonetheless, the earthquake caused an estimated 20,000 deaths, most of them in Kutch, and injured some 167,000 people. In the state capital, Ahmedabad, in eastern Gujarat – more than 250 kilometres (160 miles) from the epicentre – as many as fifty multi-storeyed buildings collapsed, notwithstanding their framed, reinforced concrete construction (although shoddy workmanship was partly to blame). In Kutch, four major towns – Bhuj, Bhachau, Anjar and Rapar – were damaged seriously enough to require the imposition of official emergency measures; the main hospital in Bhuj was completely demolished; 178 villages totally collapsed, and 165 others suffered more than 70 per cent destruction. In total, 783,000 buildings were damaged and 339,000 others destroyed, three-quarters of them in Kutch, representing 90 per cent of the district's housing stock. In Bhuj there was an oil leak; at the port of Kandla, highly toxic chemicals leaked into the air; while at

Navlakhi, a port on the Gulf of Kutch, coal dust and fluorspar spilled into the intertidal waters.

Geologically speaking, the 2001 Gujarat earthquake 'qualifies as the most devastating intraplate earthquake in the world', notes the *Bulletin of the Seismological Society of America*.[1] In other words, it occurred not at a plate boundary, like the San Francisco earthquake of 1906, but rather within a tectonic plate, like the New Madrid earthquakes in 1811–12. Gujarat lies about 400 kilometres (250 miles) from the boundary between the Indian plate and the Eurasian plate (which includes Kashmir, where a major interplate earthquake in 2005 claimed at least 86,000 lives). As with most earthquakes in the Indian subcontinent, the 2001 earthquake was caused by the build-up of stress generated by the Indian plate pushing northwards into the Eurasian plate at the rate of about 29 millimetres (1.15 inches) per year according to Global Positioning System (GPS) studies, as part of the complex process that originated the upthrust of the Himalayas 40–50 million years ago.

Gujarat, and neighbouring Sindh, therefore have a lengthy, if largely unremembered, history of earthquakes. As noted by three geophysicists in 2010:

> geological forces, particularly earthquakes, have greatly altered the geography of the region over the past 4000–5000 years or so. As a consequence, this may have accelerated the demise of many ancient settlements by altering the water supply, modifying trade routes, producing the need for continual rebuilding, and ultimately forcing migration.[2]

Archaeological evidence from the Indus civilization, which flourished in the Indus valley and in Gujarat between 2500 and 1900 BC, suggests that earthquakes may have occurred fairly often at this time. The Indus civilization site of Dholavira, in eastern Kutch, shows clear signs of earthquake damage, as noted earlier. Earthquakes could have been one factor in the mysterious decline of the civilization, although the evidence is insufficient for certainty.

This conclusion would be consistent with historical evidence for an earthquake in Kutch recorded by British observers during the colonial period.

The first ruler of Bhuj is legendarily said to have founded the city in 1549 by pinioning the tail of an underground snake with a stake driven into the ground, in an unsuccessful attempt to prevent the snake from writhing and causing the ground to shake – possibly a reference to local earthquakes (though none is actually recorded). In 1815, the then ruler was deposed by the British, and an infant installed under British regency. Four years later, in 1819, a widely recorded major earthquake centred in the Rann of Kutch helped the British to consolidate their conquest of the region, providing yet another example of the coincidence of human and natural agency. A surgeon who worked in Bhuj in the 1820s, James Burnes, offered the following personal interpretation of events in his sketch of the history of Kutch published in 1839:

> The tyranny and injustice of Rao Bharmuljee had scarcely been crushed, and a new and better order of things introduced through the means of the British government, when the hand of Providence seemed to join in depriving Cutch of some of the instruments of cruelty. A violent shock of an earthquake, attended with some extraordinary circumstance, levelled with the dust nearly all the walled towns in the country, and anticipated an intention, which had often been conceived, of dismantling some of these nests of discontent and treason. The desolation which ensued can scarcely be imagined. In Bhooj [Bhuj] alone, seven thousand houses were rent to their foundations, and twelve hundred persons buried in the ruins. Anjar suffered equally in proportion, and much injury was sustained, with the loss of many lives, at Mandavie and other large towns.[3]

In a repeat performance during the earthquake in 2001, roughly hewn stone forming the outer layer of the bastion of the inner fortress in Bhuj – apparently repairs effected after the 1819 earthquake – fell away to reveal an older façade elegantly ornamented with marching elephants that dated from the pre-colonial period.

Furthermore, the 1819 fault movement threw up a natural dam, known as the Allah Bund – that is, the 'Dam of God' (as opposed to a man-made dam) – which has been much studied by geologists and seismologists ever since. The Allah Bund was initially a ridge about 6 metres (20 feet) in height and some 6 kilometres (3.5 miles) wide, extending for at least 80 kilometres, possibly as far as 150 kilometres (50–90 miles). Its effects on the water supply of Kutch, and the area's desertification, are not certain, according to scientists. Nevertheless, the natural dam is widely believed by the people of Kutch to have diverted the waters of the Indus delta, causing fertile lands to wither, observes anthropologist Edward Simpson in *The Political Biography of an Earthquake*, his intimately informed study of the 2001 earthquake based on a decade of field research in Kutch, both before and after the disaster. 'As agricultural lands withered, the population of Kutch turned to trade, commerce and international migration for its fortune. The earthquake produced a new kind of people and society', Simpson concludes.[4] Thus, the 1819 earthquake was a significant factor in generating the Gujarati diaspora of the 19th century, initially around the Indian Ocean, including colonial Bombay and Africa. This in due course established the commercial reputation of Gujaratis overseas evident in today's Britain – many of whom donated handsomely to the reconstruction efforts in Kutch after the 2001 disaster through caste and temple associations, trusts, charities and right-wing political organizations, such as the Vishwa Hindu Parishad (VHP) and the Rashtriya Swayamsevak Sangh (RSS). According to the family lore of these migrants, says Simpson, 'the earthquake raised a dam; in turn, the dam gave rise to a diaspora.'[5]

Between 1819 and 2001, there was one other notable earthquake in Kutch. In 1956, the town of Anjar was badly damaged by a magnitude-6.1 quake, which killed more than a hundred people. By then, of course, British colonial rule was over, and India was governed by the Congress Party, led by Jawaharlal Nehru. The prime minister took a personal interest in the disaster, and paid a visit to Gujarat, the birthplace of his late mentor, Mahatma Gandhi, whose ashram was located near Ahmedabad. Speaking in Bhuj, Nehru told a large crowd: 'The deaths and destruction [are] a matter of great sorrow. However, the sufferers should rebuild the devastated areas themselves and in a better way. Make them better places to live in.'[6] Nehru inaugurated a rehabilitation centre in a village, which was renamed Jawaharnagar in his honour. At the same time, plans were laid for the building of a new Anjar, west of the old town, with a foundation stone laid by Nehru.

But, as so often with earthquakes, building and rebuilding are guided by sentiment, history and economics more than by rationality, science and engineering. In both 1819 and 1956, it was the same area of Anjar that collapsed, remarked the chairman of the Indian Institute of Engineers after the second earthquake. A minister in the state government responded that the rebuilt houses in this area would be 'quake-proof'.[7] The area went on to become the main bazaar in Anjar post-1956. On Republic Day in 2001, the area's buildings again collapsed, crushing a street parade of almost 200 children waving the national flag. The village of Jawaharnagar collapsed, too. This time around, reconstruction was carried out in the name of the current leader of the Congress Party, Sonia Gandhi, who had married into the Nehru–Gandhi dynasty. When Simpson returned to Jawaharnagar over the years following the earthquake, he found that 'the rubble had been cleared, a new village planned and constructed, promises had been made for a quake-proof and prosperous future', and a new memorial plaque had been installed next to the original plaque about Nehru.[8] There was little awareness among the villagers, and no irony, about the way in which incompetent Congress policies had encouraged seismic history to repeat itself.

By 2001, the Congress Party had been ousted from power at the centre by a national coalition government led by the Hindu nationalist Bharatiya Janata Party (BJP), elected in 1998. Gujarat was a crucial state for the BJP. It was from Gujarat that the nascent political party had launched a religious stunt in 1990 that led to the hugely controversial demolition of a Mughal-period mosque in Ayodhya in 1992, widespread Hindu–Muslim violence and the arrival of the party as a national political force. Since the 1998 elections, Gujarat had been ruled by the BJP under a chief minister, Keshubhai Patel, who was a party stalwart close to the militant far-right organization, the RSS. In 1948, Gandhi had been assassinated by a member of the RSS; by now, Gandhi's home state had moved far away from his core beliefs.

But it was not Patel's political associations that caused his downfall as chief minister in 2001; many in the BJP leadership had close links with the RSS, the VHP and other Hindu nationalist organizations. The chief cause was the earthquake and its destruction of much of western Gujarat. After the quake struck in January, the chief minister was rapidly accused of incompetence in his response by many of the survivors. In Bhuj, demonstrations persistently demanded his resignation. At the same time, his cabinet was the target of corruption allegations. When the BJP prime minister, Atal Bihari Vajpayee, and his deputy came from New Delhi to visit the ruins of Bhuj in June 2001, they were greeted with a strike and black flags flying in the streets. Then, the BJP lost a key local election in Gujarat. In October, the national leadership decided to replace Patel with a capable local organizer who had also served as the BJP's national secretary: Narendra Modi. Besides never having stood for election outside his own party, Modi was the first RSS *pracharak* (volunteer to the cause) to become a chief minister in India. Without the happenstance of the earthquake, it is inconceivable that Modi would have stepped into Gujarat's top political job, and hence into Indian national politics.

To begin with, Modi's focus was far from the earthquake and its aftermath. In February 2002, Hindu–Muslim riots broke out in Gujarat, after the burning of a train containing Hindu pilgrims returning from Ayodhya led to a state-wide pogrom against Muslims. The state apparatus proved to be 'somewhere between supine and complicit in the bloodshed', in the view of Simpson and the majority of commentators.[9] In July, Modi resigned as chief minister and the state assembly was dissolved. But in December, after an election in which the BJP, supported by the RSS and the VHP, gained a two-thirds majority by mobilizing the electorate on religious lines, Modi was again sworn in as chief minister. For the next decade, however – during which a Congress-led coalition prevailed in the centre after the BJP's defeat in 2004 – Modi was subject to travel bans by the European Union, the United Kingdom and the United States, because of his role in the 2002 riots. Although he was formally cleared of complicity by the Supreme Court of India in 2012, his reputation remains indelibly stained by his handling of the riots.

When, in 2003, Modi finally turned his attention to the reconstruction of Kutch, he homed in ruthlessly on the possibilities for using the earthquake destruction to develop the district with a new infrastructure – wide roads and twenty-four-hour electricity – and to introduce new industries. He was less interested in responding to the public clamour for reconstruction of damaged or ruined towns and villages. This was left largely to others, whether in the government, non-governmental organizations, the private sector or international charities. Most of Bhuj was rebuilt between 2003 and 2005, without any of the radical demolition of buildings and relocation of the population proposed in initial plans for the city – as with most earthquake-struck cities. On the earthquake's third anniversary, in 2004, speaking at the inauguration ceremony of two reconstruction projects in Bhuj – where more than 2,000 people had been killed by collapsing buildings – an impatient Modi went so far as tell the thousands attending his speech that they should try to forget about the earthquake, as the rest of Gujarat – that is, the developed eastern part with the state capital – already had. There is not a single reference to Chief Minister Modi in a history of Bhuj and its reconstruction by Azhar Tyabji published in 2006.

Under Modi's chief ministership, Kutch experienced a spate of industrialization along the lines of the industrial development of China since the 1980s. Instead of the open countryside visible before 2001, large areas around the Gulf of Kutch now consist of factories with smoking chimneys, often without any tell-tale logo – including the world's largest oil refinery. According to Modi, the 'backward region' demolished by the earthquake in 2001 has come to resemble Singapore, as celebrated in a 2011 publication, *Love Gujarat: A Land of Will and Wisdom*, sponsored by some leading business figures of the Gujarati diaspora in London. 'In the past, similar productions have contained images of temples, palaces and other cultural glories', notes Simpson. Now, 'The land in which Gandhi was born and developed his ideas, the "land of will and wisdom", has become a land of "unstoppable growth", "development" and smoke stacks.' During the decade of his research, he observes, 'the entire language of public politics in Gujarat . . . altered, and, in no small part, I think the destruction of the earthquake was one of the principal catalysts or "enablers" of this shift, from the watchword "Muslim" to those of "development" and "growth".'[10] Although Simpson accepts that much of this industrialization might have occurred without the earthquake, he is convinced that its physical destructiveness facilitated the economic 'earthquake' that followed it.

This so-called 'Gujarat model' of development was constantly discussed in the rest of India in the lead-up to the country's national elections in 2014. More than any other plank of Modi's election platform as leader of the BJP, his promise of economic growth, based on his track record in his home state, was what led to the victory of the BJP and his election as prime minister.

'Vote for the BJP, the argument goes,' wrote a reporter in the London-based *Financial Times* three months before the elections, 'and Mr Modi will do for the rest of India what he has done for Gujarat during his twelve years as chief minister: encourage investment, improve roads, electricity and water supply, and create the jobs desperately needed by the 10–12 million young Indians entering the workforce each year.'[11]

However, the promise rested on a questionable premise. While there can be little doubt that the economy of Gujarat did grow faster than that of most other Indian states in the early years of the new century – more than 10 per cent annually on average in the period 2006–12 – what proportion of this growth should be attributed to Modi's leadership, as opposed to the state's long history of success in industry and international trade? Ahmedabad, for example, was long known as the 'Manchester of India' for its textile mills, established in the 1860s; since 1947 it has hosted India's largest association for textile research. In the telling words of an operations manager of one of the oil refineries on the Gulf of Kutch, completed in 2006: 'Gujarat has always been a major industrial hub, and the attitude of the government has always been to invite industrial activity.'[12]

Two UK-based academic economists interested in Modi's claim to be the driving force of this growth, Maitreesh Ghatak and Sanchari Roy, noted in a paper published in early 2014 that:

> there is very little systematic evidence to evaluate Gujarat's performance under Modi [against] its own past growth record, *and* [against] that of the rest of the country or other states. This is unsatisfactory, given that the standard research method in this context is the 'difference-in-differences' approach: did Gujarat's growth under Modi compared to its growth in the previous period increase by a higher margin than the corresponding figure for the whole country or other states? Just the fact that Gujarat had a higher rate of growth than the whole country during the period Modi was chief minister is not considered good enough evidence in favour of a 'Modi effect' on growth. The difference between Gujarat's growth rate and that of the whole country during Modi's rule has to be significantly higher than what it was in the earlier period for such a claim to be made.[13]

Having carefully analysed the available data on income and output in all Indian states from the 1980s to 2010, Ghatak and Roy concluded:

> Gujarat's growth rate was similar to or above the national average in the 1980s, depending on the method of calculating the growth rates. Also, there is definitely evidence of growth acceleration in Gujarat in the 1990s, but there is no evidence of any *differential* acceleration in the 2000s, when Modi was in power, relative to the 1990s, both with respect to the country as a whole, as well as other major states. This is robust to using alternative measures of income, alternative methods of computing growth rates, and keeping or dropping the year 2000–01, for which Gujarat had a negative growth rate due to the earthquake.
>
> So the Gujarat growth story in the last two decades is definitely real and worthy of attention. However, using the difference-in-differences approach, we do not find any evidence in favour of the hypothesis that Modi's economic leadership has had any significant additional effect on its growth rate in the 2000s.[14]

Whatever or whoever is deemed responsible for the recent economic growth in Gujarat – whether its long history of commercial success, or the earthquake disaster of 2001, or its chief minister Modi, or a combination of all three – there can be no question that the benefits of this growth have not been widely shared among the people of the state. The government's focus on industrial growth led to a relative decline in investment in basic services for the poor. Little attempt was made to relate industrialization to social life. Whereas the state's per capita income is high by Indian standards, ranking eighth out of more than thirty Indian states and territories, its social indicators are not impressive. Basic literacy in Gujarat, according to the Indian national census of 2011, is only 79 per cent, well below the most literate state, Kerala, with 93 per cent. Health in Gujarat is poor, with 44.6 per cent of children below the age of five suffering from malnutrition and 70 per cent of children suffering from anaemia, according to the United Nations' *Human Development Report*. The ratio of girls to boys, 886 out of 1,000 – an implied measure of female foeticide – is seven places from the bottom of the table for all Indian states and territories. Gandhi, if he could see his birthplace now, would surely be horrified.

Looking back over the decade after the earthquake, Simpson described the changing atmosphere in Kutch in an interview with an Indian national newspaper, *The Hindu*, as follows:

> The rate of growth in some parts of the region affected by the earthquake was tremendous. A frontier mentality emerged, and industry of various kinds appeared across the region. Factories were put up at a tremendous rate. People descended on the region to profiteer . . . On the whole people I know well in Gujarat are generally much better off now than they were before. How much of that would have happened anyway [without the earthquake], is impossible to work out. But growth for growth's sake, the saying goes, has the same mentality as a cancer . . . I sensed a great deal of excitement in the aftermath of the earthquake about the coming of industry, [but] towards the end of this research I also sensed something of a sadness at the realisation that industry had changed the world and the changes were not always good.[15]

In such an atmosphere, corruption inevitably flourished. The president of the Bhuj chamber of commerce at the time of the earthquake was arrested in 2009, but later released. In 2011, just before the tenth anniversary of the quake, he ingested rat poison and died in bed, next to his sleeping wife. 'In this miserable and memorable act, he became the earthquake's latest victim. He will not be the last.'[16]

Perhaps needless to say, none of the reservations of academic experts – whether economists, anthropologists, city planners or others – about growth in Gujarat when Modi was chief minister exerted much effect on the Indian electorate in May 2014. Modi shrewdly launched his election campaign by giving an Independence Day speech in August 2013 not from the expected New Delhi or from Gujarat's capital at Ahmedabad, but rather from the grounds of a reconstructed college in Bhuj, near the epicentre of the earthquake. Here he pointedly told an audience of 25,000 people: 'I am standing here in Kutch from where my voice reaches Pakistan first and Delhi later.'[17] But in addition to its proximity to Pakistan, Modi had clearly selected Bhuj so that he might also capitalize on the reconstruction and development of Kutch after the earthquake. 'Since 2001, the devastated Bhuj has been completely rehabilitated by Narendra Modi's governance abilities', commented a leading Indian news website. 'So, Modi must have felt pride while delivering his speech from Lalan College, Bhuj. Modi's work post the Bhuj earthquake can also be contrasted with the "ruins" of the Indian economy under the Congress-led UPA government.'[18] Plainly, the website's reporter had swallowed the myth of the 'Gujarat model'; and so, most probably, did tens of millions of Indians who voted for the BJP in the elections of 2014.

Modi himself was very reticent in his direct public references to the earthquake that brought him to power in Gujarat – whether in its immediate aftermath, during the reconstruction period, during his national election campaign or after he became prime minister of India. Remarkably, there is no major statement by him about the quake cited by Simpson in *The Political Biography of an Earthquake*, nor does Modi make reference to it in a biography, *The Modi Effect* by Lance Price, written with its subject's cooperation and published in 2015. Only after a magnitude-7.8 earthquake struck Nepal in 2015 – killing about 9,000 people, burying many climbers in an avalanche on Mount Everest and collapsing tens of thousands of buildings in Kathmandu and the Kathmandu valley, including historic Hindu temples – did Modi briefly lift his public silence on the Gujarat earthquake. While expressing sympathy for Nepal and offering Indian government relief to India's stricken neighbour, he remarked: 'I saw the Bhuj earthquake of 26 January 2001, very closely.'[19]

To Modi's credit, he reacted with exceptional urgency to the Nepal earthquake, and India provided Nepal with extensive and effective relief. But it remains to be seen if this episode will translate into any substantial development of earthquake-related advance planning within India itself, where it is badly needed. The prognosis, based on the Indian response to the 2001 Gujarat earthquake, is not encouraging.

'Earthquakes are a highly underrated risk in India', writes Indian disaster management specialist Aromar Revi, in *Recovering from Earthquakes*, a collection of articles about Indian seismicity published in 2010. 'This is largely because we have been living in a "seismic gap" with no great earthquake for 75 years.'[20] Despite the occurrence of some ten major earthquakes (of magnitude 6.0 or larger) in independent India – all but one of them in the northern half of the subcontinent – there has been no great earthquake affecting northern India's chief centres of population, such as Delhi, Varanasi (Benares) and Calcutta, since the Nepal–Bihar earthquake in 1934. This had an estimated magnitude of 8.1 and an epicentre in the lesser Himalayas. It caused immense destruction across an area of Nepal and northern India spanning 465 kilometres (290 miles) from north to south and 320 kilometres (200 miles) from east to west, including Kathmandu and Calcutta, where the tower of Saint Paul's Cathedral collapsed. However, there were exceptionally few fatalities (up to 10,700), because of the quake's timing during the afternoon, when most rural inhabitants were out of doors working in the fields. The Gujarat earthquake of 2001 does not qualify as a 'great' Indian earthquake, based on its magnitude and extent of destruction – nor does the Nepal earthquake of 2015. According to seismologists' estimates of stress in the Indian–Eurasian plates, an Indian 'Big One' is still to come: 'a large future earthquake lurks in western Nepal,' notes Roger Bilham.[21]

If and when this does happen, disaster on a grand scale is very likely, warns Revi, for the following reasons:

> India's stock of over 240 million buildings is slowly transforming itself in response to rising per capita incomes and savings, especially in urban areas. . . . In metropolitan centres, where a third or more of the population lives in informal settlements in cities like Mumbai or Delhi, millions are at risk from ground shaking, [soil] liquefaction and the fires that often accompany a major earthquake in hyper-dense settlements. Many of the multi-storey framed RCC [reinforced cement concrete] buildings in so-called modern materials perform not very differently from non-engineered, artisan-built buildings because of poor material quality, workmanship and seismic detailing

– as demonstrated by the collapse of some high-rise buildings in Ahmedabad in 2001.[22] The effect on similar multi-storey structures precariously built on mountainsides in the Himalayan foothills is horrifying to contemplate. Moreover, Indian Railways are probably at serious risk, 'because a high proportion of its bridges and culverts are well beyond their service life and are not designed according to current seismic standards', notes Revi.[23]

After the Nepal earthquake in 2015, a Delhi-based architect, Gautam Bhatia, tried to project a future great earthquake's effect on Delhi and surrounding areas, in words only too reminiscent of the disaster in Tangshan in 1976 – with the added risk of fire, as in Tokyo in 1923:

> The 7.9-[magnitude] earthquake struck at 2 a.m. when the city was asleep. It lasted barely 20 seconds, but its destructive power flattened almost 60 per cent of the capital, killing half the city's 16 million population. First to collapse were the illegal colonies of East Delhi and the trans-Jamuna [an area built on the bed of the Jamuna River] and the multiple-builder high-rises in Noida and Gurgaon, many built with local sanction but cleared without the requisite structural inputs for earthquake protection. With little preparedness and inadequate machines and supplies, disaster teams were slow to react; as fires from gas leaks raged across the city, hospitals were ill equipped to deal with the injured and homeless, who spent the night huddled in parks. Uber cabs were seen ferrying dead bodies for mass cremation in what was once Connaught Place . . .[24]

As both Revi and Bhatia disturbingly conclude, there has been little pre-emptive planning against Indian earthquakes since the disaster in Gujarat in 2001, other than the establishment of a National Disaster Management Authority in 2001 and the seismic retrofitting of a few public buildings in Gujarat and heritage buildings in Delhi. When, in 2005, the Congress-led government approved a US-backed project to retrofit some of its own official buildings in New Delhi, including the secretariat and police headquarters, the project was left unfinished as the city authorities turned their attention to construction for the Commonwealth Games in 2010. At the same time, the Indian government persistently refuses to share its seismic data with the world – unlike the Chinese government. The chief stumbling block to change in India is a lack of perceived urgency, in the absence of any actual earthquake disaster, coupled with the enormous potential public and private expense of seismic design and retrofitting, not to mention the size and diversity of India. In the meantime, northern India's dense urban population can only hope, and pray, that the 'seismic gap', lasting for the past eight decades since the 1934 earthquake, will persist.

CHAPTER

10 WAR AND PEACE BY TSUNAMI: THE INDIAN OCEAN, 2004

One of many boats caught by the Indian Ocean tsunami in 2004 and deposited inland, in this case on top of a house in Banda Aceh, northern Sumatra.

Although India has yet to experience another great earthquake since the one in 1934, the Indian subcontinent, including Sri Lanka, suffered serious losses from the great Sumatra–Andaman earthquake in 2004. In fact, the deaths in India in 2004 were twice those in the Nepal earthquake in 2015. They reached almost the mortality of the Gujarat earthquake in 2001.

At first glance, this high death toll seems puzzling because the earthquake in 2004 occurred far away from the Himalayas and not even on the Indian mainland. Its epicentre was nowhere near the subcontinent. It was located on the Indonesian side of the Indian Ocean, some 160 kilometres (100 miles) west of the northern end of the island of Sumatra, roughly south of the Andaman Islands, at a depth of 30 kilometres (19 miles) below mean sea level.

One reason for the earthquake's high impact on India was its size. It lasted for over ten minutes: a record for the duration of any earthquake. Its magnitude was initially estimated at 9.0, but soon upgraded to 9.3: almost as great as the largest earthquake ever recorded, 9.5, in Chile in 1960. Although the magnitude was subsequently downgraded to 9.1, the Sumatra–Andaman earthquake is still the third largest quake, after the Chile quake and a magnitude-9.2 earthquake in Alaska in 1964, since the invention of the seismograph in the late 19th century. Yet as one of the seismologists who proposed the upgrade to 9.3, Seth Stein, noted: 'the number itself wasn't important. The important fact was that the area that had slipped was three times bigger than first thought.'[1] An area off Sumatra of staggering dimensions, about 1,200 kilometres long and 200 kilometres wide (750 by 125 miles) – similar in size to half of California – was found to have slipped about 10 metres (33 feet). That is more than thirty times the size of the area that slipped during the San Francisco earthquake in 1906 and a slippage almost three times further than that of the San Andreas fault, as the Indian tectonic plate subducted beneath the Burma microplate (a part of the Eurasian plate).

But of course the chief reason for the heavy death toll in India was the tsunami in the Indian Ocean generated by this plate-tectonic upheaval in the seabed. A tsunami of comparable size was generated by the giant volcanic eruption on the island of Krakatoa, near the southern tip of Sumatra, in 1883. Most of Krakatoa's victims died not in the eruption but in the tsunami. Krakatoa's tsunami washed away 165 villages in the Sunda Strait between

Java and Sumatra, drowning more than 36,000 people in the local region alone. Its waves reached South Africa and even as far as the English Channel. However, the tsunami produced by the Sumatra–Andaman earthquake was far more devastating than Krakatoa's tsunami, and even more far-reaching. The largest waves reached a height of 34.9 metres (115 feet). The tsunami hit Sumatra fifteen minutes after the quake, the Andaman Islands within thirty minutes, southern Thailand within ninety minutes, eastern Sri Lanka and the east coast of India after two hours, the Maldives after three and a half hours and Kenya/Tanzania on the east coast of Africa after eight or nine hours. In the north Atlantic and north Pacific Oceans, the maximum waves from the tsunami arrived up to a whole day after the earthquake. About 230,000 people died in the countries bordering the Indian Ocean.

Most striking to seismologists was the highly directional nature of the waves. This was dictated by 'the focusing configuration of the source region and the waveguide structure of mid-ocean ridges', a group of scientists reported in 2005, after they had compared sea-level data during the tsunami obtained from tide-gauges located around the world with altimetric data from space satellites, including two satellites that serendipitously passed over the Indian Ocean about 150 kilometres (90 miles) apart some two hours after the quake.[2] The fault in the source region was oriented nearly north–south. Hence, the vertical component of its slippage during subduction caused the tsunami to propagate most intensely to the east and to the west – that is, towards Sumatra/Indonesia and Sri Lanka/India, respectively – not so much to the north and to the south. On the seabed, the mid-ocean ridges such as the Southwest Indian Ridge and the Mid-Atlantic Ridge channelled the waves south of Africa and into the north Atlantic, while the Southeast Indian Ridge south of Australia, the Pacific-Antarctic Ridge and the East Pacific Rise channelled them into the east Pacific. So, Burma (Myanmar) and Bangladesh, to the north of the epicentre, largely escaped the death and devastation inflicted on Sumatra, Sri Lanka and India, as well as Thailand, some way to the northeast of the epicentre. And the Cocos Islands, a mere 1,700 kilometres (1,050 miles) almost due south of the epicentre, recorded lower wave amplitudes from the tsunami than Nova Scotia in Canada and the coast of Peru on the other side of the planet!

The country worst affected was Indonesia, especially the province of Aceh at the northern end of Sumatra, where almost 170,000 people were confirmed dead or found to be missing. Here the waves reached their maximum height. Sri Lanka was next in line, with losses of more than 35,000 people. India – where the tsunami's waves unfortunately coincided with high tide on most of the country's east coast – lost about 18,000 people. In Thailand, more than 8,000 people died, including hundreds of foreign tourists spending the Christmas holidays on the beach. Kenya/Tanzania, by contrast, suffered only eleven deaths, and Bangladesh only two deaths. In addition, half a million people were displaced in Indonesia, a similar number in Sri Lanka and some 650,000 people in India. In total, approximately 1.7 million people were permanently or temporarily displaced by the disaster.

Facts and figures aside, the reach and power of the tsunami can be felt in its impact on the southeast Indian state of Tamil Nadu, which includes the southernmost tip of India, Kanyakumari. There, tourists and pilgrims were visiting by boat a celebrated rock memorial to the Hindu monk, Swami Vivekananda, and on a neighbouring rock a 40-metre (130-foot) statue of Thiruvalluvar, a Tamil poet and philosopher. Geologist Ted Nield describes the unique and alarming scene on the morning of 26 December 2004 in his book, *Supercontinent*:

> The tourists and pilgrims, now disembarked at the Vivekananda Memorial, watched as the horror unfolded. The morning was calm and the sky clear and blue. The first thing the visitors noticed was the withdrawing roar of a false tide, as though someone had pulled a plug on the ocean. Dark, wet rocks at the foot of the many islets, and finally the seabed in between, were suddenly exposed. It was as though the sea had inhaled. In the eerie quiet, which had almost silenced the chatter on the Memorial, the visitors could hear the hiss of air being sucked into the pore spaces of the draining sand and the flapping of a few stranded fish. Then, just as the onlookers had begun to shrug their shoulders at the sight, a series of huge waves, each several metres high, rushed in, crashing over the sunstruck promenades surrounding the Vivekananda Memorial. The great statue of Thiruvalluvar was engulfed in spray, like a deep-sea light breasting an Atlantic storm, but on a cloudless morning.[3]

The visitors were unharmed by the pouncing waters. But in the district of Kanyakumari as a whole, 808 deaths were reported. Fishermen and villages further up the southeast coast, unprotected from the Indian Ocean by the island of Sri Lanka, were still less fortunate: their houses were smashed and thousands of people were swept away. One Tamil village, Nallavadu, in the small territory of Puducherry (Pondicherry), had a lucky escape, however, thanks to a fisher family's quick thinking. When the earthquake struck, the son of the family was in Singapore, and saw a television news report about the quake and giant waves said to be spreading across the Indian Ocean. He immediately telephoned home and told his sister to spread the word, abandon their house and move to high ground. The villagers swiftly broke into a small local research centre set up by some scientists to provide satellite-based weather forecasts for the Bay of Bengal, and used its public-alert system to warn the 500 families in the village to run for their lives. As a result, every single person in Nallavadu's population of more than 3,500 survived the tsunami, which reduced their village of 150 houses and 200 boats to rubble and matchwood: one of almost 400 villages thus affected in Tamil Nadu and Puducherry.

In Sri Lanka, there was no warning for coastal dwellers, and virtually zero awareness of tsunamis. No historic record of a Sri Lankan tsunami exists, unless an account in the island's ancient chronicle, the *Mahavamsa*, of what appears to have been a tsunami – in which the sea rose and flooded part of the kingdom ruled by Kelanitissa in the 2nd century BC – is accepted as fact. A Sri Lankan-born academic who only narrowly survived the tsunami, Sonali Deraniyagala, admits in her memoir, *Wave*: 'This was the first time I'd ever heard the word.'[4] A post-tsunami joke in Sri Lanka had it that a minor official in the prime minister's office received a phone call warning that 'Tsunami' was on the way from Indonesia, and despatched a welcoming party to Colombo airport to greet the unknown Indonesian guest.[5]

The tsunami affected over 70 per cent of Sri Lanka's coastline, stretching over 1,000 kilometres (650 miles), from Jaffna in the north down the eastern and southern coast and as far up the western coast as Chilaw, north of Colombo. About 88,000 homes and 24,000 boats (including 75 per cent of the fishing fleet) were destroyed, and coastal infrastructure was severely damaged. A crowded train operating on the coastal line from Colombo south to Galle was first flooded by a tsunami wave 200 metres (220 yards) from the sea at the village of Peraliya, and then picked up and smashed by a second wave, killing at least 1,500 passengers, probably more: a world record for a single rail disaster. That said, damage on the southwestern and southern coasts was relatively moderate because of the protection afforded by headlands, dunes and steep coastal gradients, except in Galle and Hambantota, as compared with the more exposed northern and eastern coasts. The latter areas accounted for two-thirds of Sri Lanka's tsunami deaths and 60 per cent of the displaced; half of these deaths occurred in one eastern district, Ampara, which meets the Indian Ocean in a large strip of flat land. By a fateful coincidence, these northern and eastern provinces were also those most affected by a long-running military conflict, between the Liberation Tigers of Tamil Eelam and the Sri Lankan government.

Near the earthquake's epicentre, in northern Sumatra, the interior of Aceh was protected by being mountainous. But in the coastal plains that surround the interior the waves did immense damage and offered little chance of escape. They were naturally highest on the western side, facing the epicentre, reaching over 30 metres (100 feet) and overtopping the mosque in a village at the tip of Sumatra, just 10 kilometres (6 miles) from the provincial capital, Banda Aceh. There, in the 'Port of Aceh', they peaked at around 9 metres (30 feet) in the shallow coastal waters of the Andaman Sea.

Great earthquakes, often accompanied by fires, produce a cityscape akin to heavy aerial bombing, as we know from earthquake-struck cities such as Lisbon, San Francisco and Tokyo. The effect of a great tsunami is less familiar, more surreal. In Banda Aceh, the waves stranded huge ships and fishing boats kilometres inland and swept cars into the sea, while leaving vast open stretches of flattened housing, with here and there a more sturdily built construction left standing. Many mosques in Aceh survived, though damaged. The pious regarded this as a sign of God's anger at the senselessness of another long-running military conflict, between the Free Aceh Movement and the Indonesian government; non-believers attributed the mosques' survival to their construction either by Dutch engineers during the first half of the 20th century or by modern contractors who were less inclined to cut corners when building a mosque than when building a school or housing complex. Indeed, the 'miracle' of the mosques became a factor in the upsurge of Islam during the rebuilding of Aceh, post-tsunami.

One survivor, a local woman, inadvertently shared the horror of her experience with a foreign visitor and writer who had recently served as a Reuters correspondent in Indonesia, Elizabeth Pisani. In the aftermath of the tsunami, this Acehnese woman had deliberately moved away from the coast to the highlands, and was working as a receptionist at a hotel where Pisani happened to be staying in April 2012, when Indonesian television news announced an 8.6-magnitude earthquake near Aceh and predicted another tsunami soon to follow. As panic took hold in Aceh on the TV screen, the chandelier in the ceiling of the hotel's lobby started to tremble and then shake. Everyone quickly stepped outside into the cold mountain rain. There, a second, apparently bigger, quake made Pisani's teeth rattle along with the hotel's windows. Her arm was now gripped by the receptionist so hard that her hand went numb. The blood had drained from the woman's face, and her knees were weak, but she would not sit down, so as to avoid feeling the shaking more, nor would she go back inside the hotel. She would only stand in the rain and keep whispering one word: 'Trauma. Trauma.'[6] Later, the receptionist told Pisani that her mother, her brother and two sisters had been swept away by the tsunami in 2004. Fortunately, in 2012, the predicted tsunami did not materialize.

Saiful Mahdi, currently director of the International Centre for Aceh and Indian Ocean Studies in Banda Aceh, lost sixteen members of his family in Aceh to the waves in 2004. He himself was then working in the United States. Looking back in a film documentary made for the tenth anniversary of the tragedy, Mahdi remarks with stoic dignity: 'Recently my professor from Cornell University told me, a flood might only affect the poor, but a tsunami will affect the poor, the rich, the powerful, the military, ordinary people and even the most powerful person. That's why a tsunami is the most democratic disaster in history.'[7]

The overwhelming majority of the dead bodies that littered the coastlines of the affected countries had to be buried in mass graves, for lack of identification. Only 5,000–6,000 of the tsunami's fatalities were formally identified, most of whom were foreigners. This was especially true in Aceh, where the ratio of dead to injured was 6 to 1, as compared with 1.5 to 1 in Sri Lanka, due to Aceh's proximity to the epicentre, its coastal population density and the poor quality of its housing, plus the social disruption caused by the province's war, which had displaced more than 300,000 people in the years before the natural disaster. So stunning was Aceh's trauma that there is 'almost no death' on display in today's tsunami museum at Banda Aceh, observes Pisani. It is a 'monument to something you don't really want to help people to remember. A museum of amnesia.'[8]

The international response to the catastrophe was immediate and generous – not least because of the widespread media coverage of the deaths and injuries of foreign nationals, especially in Thailand, and the timing of the disaster during the Christmas period. In fact, there developed a degree of 'competitive compassion' among foreign governments, in the words of two US-based academics, who report that:

> In the days and weeks following the tsunami, all the major powers with geostrategic interests in Southeast Asia – China, India, Japan and the United States – donated to the relief effort. Within two days of the tsunami strike, Japan announced a $30 million aid package, double the initial US pledge. The United States then raised its offer to $350 million, while China pledged $63 million. Japan responded by increasing its offer to $500 million. Thailand and India, also hit by the tsunami, asserted their regional positions by giving aid to smaller neighbours while themselves rejecting most outside assistance.[9]

Inevitably, the devastation, followed by the influx of foreign aid money and personnel, also produced calls for resolution of the military conflicts in Aceh and Sri Lanka, both nationally and internationally. But the long-term political effects of the tsunami on these two conflicts turned out to be diametrically opposed. Whereas in Aceh the rebels were rewarded with concessions by the Indonesian government, and the war ended in 2005, in Sri Lanka they were forced into a corner of the country by the Sri Lankan government and annihilated in a bitter military struggle in 2009. To understand why these two outcomes were so different, we need first to trace the historical roots of the two conflicts. Aceh's is probably simpler to follow than Sri Lanka's, so let us begin with Aceh, and then consider Sri Lanka.

Aceh had a long and independent existence before its absorption as a province of the new nation of Indonesia in 1949. In the 13th century, it became a Muslim stronghold, the first in the Indonesian archipelago; and from 1511 it was known as the Sultanate of Aceh. During the 17th century, in the reign of Sultan Iskander Muda, Aceh's influence extended over most of Sumatra and also the Malayan peninsula. Despite frequent contacts and struggles with European colonial powers – first the Portuguese, then the Dutch and then the British – the sultans of Aceh never really succumbed to colonial rule and instead built up their own control of regional trade, including more than half the world's supply of black pepper by the 1820s. But in 1873, pirates operating from Aceh became a threat to European commerce in the Strait of Malacca. Aceh was attacked by the Dutch, and after a guerrilla war lasting for more than a quarter of a century, the sultan finally surrendered in 1903 and was exiled in 1905. Even then, the Dutch were unable to pacify the Acehnese during the next few decades. Though Aceh was occupied by Japanese troops during the Second World War, the Acehnese fought back, rebelling against both the Japanese and the Dutch in 1942, under the banner of Islam. In 1946, after the defeat of Japan, when Dutch forces recaptured Indonesia, Aceh remained the only free part of the archipelago.

In 1948, the first president of Indonesia, Sukarno, promised Aceh a special provincial autonomy within the new country, with official recognition of Islam. But in reality, after 1950, Aceh was incorporated into the province of North Sumatra as part of the secular and centralized federal republic of Indonesia. A rebellion started in 1953 and lasted until 1959, when Sukarno finally granted separate provincial status to Aceh and wide-ranging freedom in regard to the practice of Islam. Relative peace prevailed until the late 1960s, when Sukarno was forced out of office by General Suharto. Then, in 1971, huge oil and natural gas reserves were discovered in north Aceh. Exploitation by the Indonesian state in collaboration with an international oil company, ExxonMobil, got underway without the involvement of the Acehnese and without benefit to Aceh. Indeed, land was confiscated from villagers without compensation by the Indonesian government, in order to build an oil refinery. All this fuelled the birth in 1976 of yet another rebellion, the Gerakan Aceh Merdeka, known as GAM or the Free Aceh Movement, which now aimed at complete independence for Aceh.

Its leader, Hasan di Tiro, was an Acehnese businessman who had failed in his bid for a pipeline contract for the new oil refinery. He was also a descendant of the last sultan of Aceh and a direct descendant of an Indonesian national hero during the war of 1873–1903, Teungku Chik di Tiro. After being wounded by the Indonesian army in an ambush in 1977, he fled and eventually settled in Sweden, along with another key leader of the movement, where they formed a government-in-exile in 1979. Di Tiro did not return to Aceh for over three decades, until after the peace agreement of 2005. Meanwhile, the Free Aceh Movement fought brutally with the Indonesian army in Aceh, especially in the 1990s. Hostilities largely ceased in 2000–3, when the Indonesian government offered Aceh concessions, notably the right to more of the revenue from natural resources, to implement sharia law and sharia courts and to create symbols of autonomous government. But when the peace talks failed in 2003, President Megawati Sukarnoputri declared martial law and a state of emergency in Aceh, and sent in large numbers of troops and police. They killed about 6,000 rebel combatants (one-quarter of the total force) and thousands of Acehnese associated with the movement. At least 125,000 people were displaced; the supporters of the movement were pushed out of the cities and villages and into the forests and mountains. By 2004, not only was the Free Aceh Movement on the defensive, it was losing the support of a large part of the Acehnese, who were desperate for peace. Recognizing this, President Yudhoyono, newly elected in October 2004, offered to reopen secret negotiations with the leaders of the movement. His offer was accepted by the leaders just four days before the tsunami struck on 26 December.

In May 2005, martial law was ended and in August, following five rounds of talks, a peace agreement known as the memorandum of understanding (MoU) was signed. It secured the withdrawal of government troops from Aceh, the disbanding of the military wing of the rebel movement, a greater share for Aceh of revenues from energy, mining, logging and fishing, greater autonomy for local government and, importantly, the right of the former leaders of the Free Aceh Movement to form local political parties. Most of the MoU's provisions passed into law in 2006, against resistance from hardliners. But despite many continuing tensions in Aceh – especially over the greater distribution of money to tsunami survivors than to war refugees – peace has prevailed since 2005.

The centrality of the tsunami was specifically noted in the memorandum as follows: 'The parties are deeply convinced that only the peaceful settlement of the conflict will enable the rebuilding of Aceh after the tsunami disaster on 26 December 2004 to progress and succeed.'[10] On the MoU's first anniversary in 2006, President Yudhoyono stated: 'The tsunami produced an overwhelming moral, political, economic, social imperative to end the conflict . . . I was criticized by those who did not see any benefit from renewed talks with GAM. But I was more concerned about the judgement of history for missing this rare window of opportunity to resolve the conflict.'[11] An ex-commander of GAM, when asked in 2007, 'What effect did the tsunami have on the peace process?' replied: 'GAM looked at the humanitarian side and decided it could not go on. The Indonesian government saw the same; regardless of [whether] they wanted to or not, they had to sign the MoU.'[12] In Pisani's view, published in 2014 in her travel book about Indonesia:

> the unimaginable tragedy of the tsunami allowed both Jakarta and the rebel leaders to climb out of the trenches they had dug for themselves and to talk peace. The torrent of support from ordinary Indonesians helped too; the rebels could no longer argue that Indonesians wanted only to take from Aceh, not to give.[13]

In summary, writes Jennifer Hyndman in her study of the tsunami and its political, economic and social aftermath: 'The tsunami did not *cause* the peace agreement, but it certainly created conditions that accelerated its signing.'[14]

In Sri Lanka, however, the roots of the civil war were far older than in Aceh. Conflict between a Sri Lankan Tamil minority, who are mostly Hindus, and the majority Sinhalese population, who are mostly Buddhists, goes back to the 2nd century BC. At this time, Elala, the Tamil ruler of the Chola kingdom in south India, also ruled the northern part of Sri Lanka, including its capital, Anuradhapura, until his defeat by the Sinhalese king, Dutthagamani, as recorded by the Sinhalese in the *Mahavamsa*. These ancient rulers remain heroes to their respective communities today.

Under British colonial rule of Ceylon (as Sri Lanka was then known), from 1815 to 1948, the Tamil population acquired a new dimension, when the colonial power imported large numbers of Tamil labourers from south India to work the coffee, tea and rubber plantations in the central highlands. By the time of political independence, 22.7 per cent of the island's population was Tamil, half of whom were Sri Lankan Tamils, the other half Indian Tamils. Moreover, the vast majority of people in the northern part of the island, around Jaffna, were Sri Lankan Tamils; they also formed a majority in the eastern part, around Trincomalee. In the rest of the island, however, Sri Lankan Tamils were a small minority.

Ethnic and religious conflict between Sinhalese and Tamils was largely suppressed until 1948. But it came into the open after independence with the government's passing of the Sinhala Only Act in 1956, which replaced English as the official language of Ceylon with Sinhala, and denied official recognition to the Tamil language. Two years later, Sinhalese nationalists provoked island-wide anti-Tamil riots, in which as many as 200 Tamils were killed. Then the government standardized admission to universities in a way that disadvantaged Tamils. In another move unfriendly to Tamils, in 1972, the country's name, Ceylon, which probably derives from the Tamil name Eelam, was formally changed to Sri Lanka, a term of Sanskrit origin meaning 'Blessed Island'. Soon after, in 1976, Tamil nationalism gave birth to the Liberation Tigers of Tamil Eelam (LTTE, often known as the Tamil Tigers), founded by the ruthless Velupillai Prabhakaran, coincidentally in the same year as GAM. As in Sumatra, so in Sri Lanka, the rebels aimed to establish a separate state in the part of the island where they were dominant.

From the first, the LTTE used extreme violence against both the government and civilians. They enjoyed considerable success, gaining control of most of the Jaffna peninsula in the mid-1980s, until the arrival of a peacekeeping force sent from India in 1987 by the government led by Rajiv Gandhi. This, however, departed the island in 1990, leaving the LTTE once again in control of the north and part of the east. Fighting continued with

Sri Lankan government forces throughout the 1990s, during which LTTE suicide bombers assassinated both Rajiv Gandhi, after he had stepped down as prime minister, and Ranasinghe Premadasa, the serving Sri Lankan president. Eventually, in 2002, after the LTTE dropped its demand for a separate state in favour of regional autonomy and after the election of a new government in Colombo, a ceasefire agreement was signed by the LTTE and the government, leading to peace talks. Despite sporadic violence, this agreement was still in force when the tsunami struck in 2004, severely damaging the areas in the north and east controlled by the LTTE.

As in Indonesia, so in Sri Lanka, a memorandum of understanding (MoU) for reconstruction was drawn up. Known as the Post-Tsunami Operational Management Structure (P-TOMS), and signed in June 2005, it was designed by international donor agencies with representatives of the Sri Lankan government, the LTTE and Muslim political parties (Muslim communities on the coast had been disproportionately damaged by the tsunami) to channel foreign aid directly to the affected areas of the island, not via the government. The LTTE supported P-TOMS, and called for as much autonomy as possible in the administration of aid. The Sinhalese nationalist party opposed it, and challenged its legitimacy in the Supreme Court. Their chief grounds were that the LTTE was a terrorist organization, not a governmental entity, and therefore could not legitimately participate in such an agreement. Moreover, the committees described by the MoU could not legally carry out their work without constitutional changes. Furthermore, international donor funds could not be controlled by non-governmental bodies, such as the World Bank. Finally, the Sinhalese nationalists argued that to give aid only to those in the tsunami disaster zone would discriminate against tsunami-affected victims outside the zone. Behind all these arguments was the deep-seated belief of the nationalists – shared by the majority of the Sinhalese population – that little aid money should be spent on areas controlled by the LTTE. Thus, P-TOMS became mired in ethnic politics. 'If P-TOMS had succeeded, it might have served as a blueprint for the constitutional change required for lasting peace in Sri Lanka', comments Hyndman. 'However, it did not. The Supreme Court largely agreed with the plaintiff, and the MoU was never adopted.'[15]

Later that year, a hardline Sinhalese president, Mahinda Rajapakse, was elected on an anti-P-TOMS platform. He promptly changed the name of the government's tsunami-response body from the Task Force to Rebuild the Nation to the more neutral Reconstruction and Development Agency. Then he began to consolidate his power base and increase the military budget in preparation for an assault on the LTTE's northern stronghold. At the same time, the LTTE escalated its military activity. In early 2008, Rajapakse's government ended the ceasefire agreement of 2002, and the war resumed, with atrocities on both sides against civilians, whom the LTTE compelled to remain in the conflict zone. During the first five months of 2009, more Sri Lankans were killed in the fighting than had died in the 2004 tsunami. Eventually, the LTTE fighters, including their leader Prabhakaran, were trapped in a small section of coastline in the northeast, where they were massacred by a government offensive in May. The war was over – but it seems unlikely that peace will prevail in Sri Lanka in the longer term, given the continuing oppression of the Tamil minority. 'The Tamil Tigers have been totally eliminated but that hasn't made the victors any more willing to share their beautiful island', concludes a former resident BBC correspondent, Frances Harrison, in her grim reportage, *Still Counting the Dead*.[16]

Why did the Indian Ocean tsunami catalyse peace in Aceh, but war in Sri Lanka? Both modern insurgencies lasted for a similar period, both enjoyed mass support, and both provoked large numbers of atrocities, deaths and enormous displacement. Yet, following the tsunami, the first war rapidly concluded in peace negotiations, the second eventually ended in slaughter.

There is no single overriding reason for the opposite outcomes. Undoubtedly, geography played a part. The location of Banda Aceh on a coastal plain near the epicentre of the earthquake resulted in far greater death and devastation – among the supporters of the Free Aceh Movement as much as the rest of the population – than in Sri Lanka's coastal regions. This compelled the leaders of the Free Aceh Movement to talk peace, but left the LTTE leadership with more room for manoeuvre. Moreover, Indonesia is a vast archipelago consisting of many large islands, whereas Sri Lanka is essentially a single island. So there was enough space in Indonesia to allow autonomy to Aceh, but not enough in Sri Lanka for the establishment of Eelam. Geography also made the distribution of aid more straightforward, and less politicized, in Aceh than in Sri Lanka. In Aceh, the chief area of military conflict was on the eastern coast not on the western and northern coasts, the area most affected by the tsunami's devastation, whereas in Sri Lanka, the areas of military conflict and of devastation by the tsunami often coincided and were controlled by the LTTE. In addition, religion acted in favour of a peace agreement with regional autonomy: despite considerable ethnic diversity, more than 98 per cent of Aceh's population are Muslims, as opposed to Sri Lanka's 70 per cent population of Sinhalese Buddhists, with significant minorities of Hindus, Muslims and Christians. So did politics. President Yudhoyono was aware that Indonesia, and his own popularity as a president, had much to gain from a lasting peace agreement in Aceh, unlike the Sinhalese nationalist leaders, led by President Rajapakse, who enjoyed widespread support in their aim of using the tsunami to weaken the LTTE. And so did history. Aceh had fought for and preserved its existence as an independent state for half a millennium before the 20th century. This historical fact was an incentive for both the Achenese and the Indonesian government to devise a mutually acceptable resolution of the contemporary conflict. By contrast, Sri Lanka had been politically unified for almost two centuries. This fact discouraged any concessions to a separatist minority.

A second positive outcome of the tsunami has been the establishment of an international warning system for the Indian Ocean. No warning system existed prior to 2004, because there did not appear to be any threat from tsunamis. No one was conscious of the tsunamis produced by Indonesian earthquakes in 1833, 1861 and 1907, nor did many people associate the eruption of Krakatoa with a tsunami. Indeed, most people in the region had never heard of this exotic Japanese word.

The Pacific Ocean, by contrast, had long had such a warning system. It was started in 1949, after an earthquake in the Aleutian Islands and a tsunami killed people in Alaska and Hawaii in 1946, and strengthened after a Pacific-wide tsunami in 1960 caused by the great Chile earthquake in 1960. Operated by the US National Oceanic and Atmospheric Administration, the warning system has centres in Alaska and Hawaii.

In the years since 2004, a network of seismographs, coastal sea-level gauges and a handful of offshore tsunameters – satellite-linked buoys in the open ocean capable of detecting the passage of large waves – has come into existence in all of the countries and major islands around and within the Indian Ocean. Ranging in location from South Africa and Oman via India and Sri Lanka to Indonesia and Australia, these instruments are monitored by three regional alert centres in Australia, Indonesia and India: the countries who funded most of the $450 million required to set up the system. When an earthquake occurs, 'Scientists there use seismic data to estimate how much the earthquake has displaced the ocean floor', notes a recent report in the journal *Nature*. 'Then they compare the real quake with model scenarios in which they have calculated what size of tsunami might be produced' – and alert national governments about what to expect on their coastlines.[17]

That's the theory, at any rate. In practice, there are many difficulties with the Indian Ocean Tsunami Warning and Mitigation System. It costs around $50–100 million a year to maintain the instruments (notably the tsunameters); and national commitment tends to be proportional to national exposure to earthquakes. Second, it leaves governments to decide exactly how to communicate the scientific warning to the communities actually at risk on the ground. Third, it is liable to false alarms, because great earthquakes do not always produce the tsunamis predicted by model scenarios. The tsunami warning after the Indian Ocean earthquake in 2012 – issued not only to Aceh but also to twenty-five countries around the Indian Ocean, along with projected tsunami landfall times as far afield as South Africa – was one such false alarm. No tsunami occurred, because the fault movement in 2012 was different from that in 2004. Rather than subduction, which causes vertical slippage, it involved mainly horizontal slippage, which displaces little of the ocean. Lastly, and perhaps most important, the warning system is not quick enough to react, given the urgent needs of coastlines close to the epicentre of an earthquake – as in Aceh in 2004 and in Japan, where Pacific tsunamis have repeatedly devastated the northeastern coast. For this reason, the Japanese depend on their own warning system, run by the Japan Meteorological Agency, not on the US-run Pacific Ocean warning system. In 2011, however, during Japan's great Tohoku earthquake, even this system would prove to be fatally inadequate.

CHAPTER

II MELTDOWN AND AFTER: FUKUSHIMA, 2011

Satellite view of the Fukushima Daiichi nuclear power plant three days after its partial destruction by a tsunami in 2011.

At 2.46 p.m. on 11 March 2011, Japan experienced the largest earthquake in its history, which shook most of the nation for about six minutes, causing the skyscrapers in Tokyo to sway back and forth like bamboo in the wind. Globally, it caused the planet to spin faster and shortened the day by a little under two microseconds. The epicentre was at a depth of 24 kilometres (15 miles), beneath the Pacific Ocean, where the Pacific plate subducts under the plate bearing the northern part of Japan's main island, Honshu, along the Japan Trench. During the earthquake, the Pacific plate shifted to the west by an extraordinary 30–60 metres (100–200 feet); compare the plate slippage of about 10 metres (33 feet) off Sumatra in 2004. 'Such huge seafloor movements have never been recorded globally', notes a leading Japanese seismologist, Kenji Satake.[1] This wrench in the seabed happened approximately 50 kilometres (30 miles) east of the Oshika peninsula in Miyagi prefecture, on the northeast coast of Honshu known as the Sanriku coast, in the region of Tohoku (which literally means 'northeast'). Often known in Japan as the Great East Japan earthquake, the quake is known internationally either as the Tohoku earthquake or as the Tohoku-oki earthquake (to signify that it occurred in the open ocean).

At 2.49 p.m., the Japan Meteorological Agency (JMA) initially advised that the earthquake's magnitude was 7.9, which was similar to the estimated magnitude of the Great Kanto earthquake in 1923. It also issued a warning that a tsunami was expected. In 2011, the JMA's tsunami warning system consisted of three categories: a tsunami advisory (estimated wave height about 0.5 metres, or 20 inches), a tsunami warning (wave height about 2 metres, or 6.5 feet) and a major tsunami warning (wave height 3 metres, 10 feet, or more). Waves higher than 2 metres can destroy timber houses, while those higher than 16 metres (52 feet) can destroy concrete structures. For the Tohoku tsunami, the JMA's estimated heights were 6 metres (20 feet) on the Miyagi coast, 3 metres on the coast of Fukushima prefecture, south of Miyagi, and 3 metres on the coast of Iwate prefecture, north of Miyagi. Many coastal residents, who were already experiencing very strong ground shaking, immediately prepared to evacuate to high ground.

Such initial forecast heights are based on the JMA's estimate of the location, depth and magnitude of an earthquake using a seismographic network consisting of several hundred stations, combined with a database of model simulations of tsunamis, so as to provide a prompt warning within a few minutes of the occurrence of an earthquake. The JMA's tsunami warning messages can then be updated, usually within fifteen minutes, on the basis of more comprehensive seismic observations – using broadband seismography, capable of detecting a wider range of vibrational frequencies – coupled with confirmation of the tsunami by satellite-linked tsunameters floating offshore and pressure gauges located on the ocean floor, thereby reducing the rate of false alarms.

However, because of the gigantic size of the Tohoku earthquake, most domestic broadband seismograms went off the scale, and so the JMA was unable to estimate its magnitude from Japanese data. As a result, the agency took more than an hour to estimate the magnitude using seismographic observations of the earthquake from outside Japan. It officially upgraded the earthquake's magnitude to 8.4 at 4.08 p.m., followed by magnitude 8.8 at 6.47 p.m.; not until 12 March, almost twenty-four hours after the earthquake, did the JMA announce its true magnitude: 9.0.

'Because the initial magnitude estimate was much smaller than the actual earthquake size, the levels of the tsunami warning and the estimated tsunami heights were underestimated', comments Satake. Adding to the lack of urgent concern was the fact that the earliest tsunami waves to arrive, at the port of Ofunato in Iwate prefecture, were only 0.2 metres (8 inches) in height, as the JMA announced at 2.59 p.m.. 'While this was a precursor to the following large tsunami, some people erroneously interpreted the initial announcement to mean that the actual tsunami was small and slowed down their evacuation.'[2]

At 3.14 p.m. – twenty-eight minutes after the earthquake, but before the main tsunami reached the coastline – the JMA, on the basis of readings from its offshore instruments, upgraded its warnings to a higher level of estimated tsunami heights: more than 10 metres (33 feet) on the Miyagi coast, and more than 6 metres (20 feet) on the Fukushima and Iwate coasts. At 3.30, the warning was further upgraded, to more than 10 metres for Fukushima and Iwate, too, and indeed for the rest of the Tohoku coast, including Chiba, a prefecture next to Tokyo. More upgrades were to follow, eventually including the entire coastline of Japan. However, the warnings did not reach all of the coastal communities, because a power failure occurred and many people had already started evacuation.

When the main tsunami actually arrived in Ofunato, the waves peaked at 23.6 metres (77 feet). This was lower than the maximum measured during the Indian Ocean tsunami in 2004, and lower than Ofunato's all-time-record height, 38.2 metres (125 feet), during the Meiji Sanriku tsunami in 1896. The latter event introduced the word *tsunami* into the English language, courtesy of an American magazine article by Lafcadio Hearn published in the same year, in which he wrote:

> From immemorial time the shores of Japan have been swept, at irregular intervals of centuries, by enormous tidal waves – tidal waves caused by earthquakes or by submarine volcanic action. These awful sudden risings of the sea are called by the Japanese *tsunami*. The last one occurred on the evening of 17 June 1896, when a wave nearly two hundred miles long struck the northeastern provinces of Miyagi, Iwate and Aomori, wrecking scores of towns and villages, ruining whole districts, and destroying nearly thirty thousand human lives.[3]

That said, the 2011 tsunami might in fact have been of higher maximum height at Ofunato than the 1896 tsunami, had breakwaters and other countermeasures against tsunamis not been adopted there since 1896 to reduce water levels.

According to historical records, Hearn underestimated tsunamis' frequency: as many as seventy have struck the Sanriku coast. Major earthquake-and-tsunami combinations occurred in 1611, 1793, 1856, 1933 and 2011. But it is clear from archaeology – palaeoseismology – that the tsunami phenomenon dates back much further. Excavations on the Sendai plain in Miyagi, which was inundated up to 5 kilometres (3 miles) from the coast in 2011, have shown multiple layers of prehistoric inundation by seawater. One earthquake in particular, the Jogan quake in 869 (also mentioned in a historical chronicle), may have been as powerful as the 2011 Tohoku earthquake. Indeed, according to Satake, such massive earthquakes may have recurred, on average, about once in every millennium.

Using historic and prehistoric dates, one can easily calculate a notional average recurrence interval for Sanriku earthquake-tsunamis. But it has no predictive value, as with the average recurrence interval of major earthquakes on the San Andreas fault. Moreover, tsunamis have unique characteristics, which also defy prediction. Not only does each depend on the nature of the fault movement of the earthquake that generates it, it is also strongly influenced by the quirks of geography, such as a coastline's bays and inlets, which can funnel waves and also sandwich people and structures between two tsunami waves coming ashore from opposite directions. In 1611, the tsunami waves (maximum height 21 metres, or 69 feet) seem to have caused more destruction as they withdrew back out to sea than in their original surge of water onto the coastline. In 1793, the wave height was low (only 4–6 metres, or 13–20 feet), but the water affected a large area. In 1856, there was strong seismic shaking, but the waves came in slowly and were relatively low (similar to those in 1793), giving people sufficient time to flee. In 1896, by contrast, the shaking was weak, but the waves came in fast and high (maximum height 38 metres, 125 feet, as already noted) – hence the great loss of life and devastation. 'The boilerplate warning for many modern financial investments that "past performance does not guarantee future results" applies even more in the case of major earthquakes and tsunamis', remarks historian Gregory Smits in his revealing study, *When the Earth Roars: Lessons from the History of Earthquakes in Japan*.[4]

The ongoing debate about the relationship between earthquakes and tsunamis began with the 1896 disaster. Between 1897 and 1900, Akitsune Imamura – the seismologist who successfully 'predicted' the Great Kanto earthquake – published his theory of the cause of tsunamis. At a time when geologists everywhere were still unsure whether faults caused earthquakes or earthquakes caused faults, the prescient Imamura was already convinced that the movement of geological faults beneath the ocean floor during earthquakes was responsible for a massive quantity of displaced water, which became a tall wave only when it eventually reached land. Hence the gentle, almost imperceptible, undulation of the wave while travelling through the ocean, and the time delay between an earthquake and the arrival of a tsunami on the coastline. Moreover, Imamura recognized that only some kinds of undersea fault movement would generate a large displacement, and that the height of a tsunami would depend on the slope and shape of a coastline.

Imamura's rival, Fusakichi Omori, rejected this theory and proposed his own in 1900. In Omori's view, each region of the world's oceans oscillated at a characteristic frequency, like a 'fluid pendulum'. This oscillation could be stimulated by waves of the same frequency, in the same way that a sound wave causes a tuning fork to resonate at a certain frequency. Seismic waves, even from a far-distant earthquake, might therefore set off an oscillation and cause a tsunami in an enclosed bay or inlet on a coastline or in a lake, as in the seiches observed in various European lakes – from Scotland to Finland – during the Lisbon earthquake in 1755. The seiche and the tsunami were similar phenomena, according to Omori, both of which could be caused by earthquakes.

When Imamura responded to Omori's theory in 1905, he focused on two weaknesses. First, if it were correct, then tsunamis should occur at almost the same time as earthquakes, given the speed of seismic waves. Yet in 1896, the Meiji Sanriku tsunami occurred thirty or forty minutes after the earthquake. Second, if Omori were right, tsunami waves should not depend on topography, only on the characteristic frequency of water in a particular ocean region. Yet in 1896, the wave heights undoubtedly had varied very significantly with the shape of a bay or an inlet on the Sanriku coastline. 'Clearly, Imamura argues, what caused the tsunami was not the water in the bays and inlets but something originating at the ocean floor offshore', notes Smits.[5] There was no response to Imamura's criticism from Omori.

Whatever the correct geological explanation of tsunamis might be, people came to understand their destructive behaviour more clearly during the 20th century. Memories of the 1896 tsunami, which claimed 22,000 lives, saved people in the Showa Sanriku tsunami in 1933, which claimed about 3,000 lives. So did stone markers erected on the Sanriku coastline with the active encouragement of Imamura, who had been personally moved by the misery produced by the 1896 tsunami and believed that the Japanese could be trained to avoid disasters. Today there are approximately 200 tsunami monuments along the Sanriku coast: 'gnarled stone tablets, the size of mini-tombstones, warning future generations to build their houses further from the shore', writes David Pilling, a former foreign correspondent in Japan.[6] Most of these monuments indicate both the 1896 and 1933 water levels and offer basic advice. The most famous, built in Aneyoshi, a village destroyed in 1933 that is now part of the city of Miyako in Iwate, carries the following inscription (in Japanese):

> A dwelling on high ground – the happiness of descendants
> Recalling the catastrophe of great tsunamis
> Do not build houses below this point
> Tsunamis in 1896 and 1933 reached this point
> The village was destroyed and survivors numbered a mere two and four people respectively
> No matter how many years pass, be on guard.[7]

The advice seems to have worked in Aneyoshi. In 2011, some three generations later, no one had built below the marker. When the Tohoku tsunami water levels more or less repeated the levels of 1896, they damaged only the ground below the marker, leaving broken and uprooted trees, but did not reach the handful of Aneyoshi's homes beyond it.

By contrast, many villages and towns on the Japanese coastline tried to resist future tsunamis through building sea walls up to 10 metres (33 feet) in height. In 2011, a third of the country's coastline lay behind walls that together are longer than the Great Wall of China. At Taro, a village on the Iwate coast destroyed in the 1611, 1896 and 1933 tsunamis, the people built a wall encircling the village, which by the late 1950s was 1,350 metres (1,500 yards) in length and rose 10 metres (33 feet) above sea level. It saved the place during the Pacific-wide tsunami after the great Chile earthquake in 1960 (magnitude 9.5); and people came from all over the world to learn from Taro's experience. Locals cycled, walked and jogged along the top of their wall. After its further reinforcement, on the seventieth anniversary of the 1933 tsunami in 2003, Taro publicly declared itself to be a 'tsunami defence town'. The tsunami from a second great Chile earthquake in 2010 (magnitude 8.8) reinforced this over-confidence: predicted by the JMA to have a maximum wave height of 3 metres (10 feet), it reached less than 2 metres (6.5 feet). 'But a tsunami arising from a giant earthquake on the other side of the planet and a tsunami from a giant earthquake only a short distance off your own coast are quite different', remarks seismologist Roger Musson.[8] In 2011, Taro was once again wiped out by a tsunami, which easily overtopped the famous wall and even ruptured it in several places. The same unstoppable flooding occurred in many other coastal villages and towns. In Rikuzentakata, a town of more than 23,000 people in Iwate, also devastated in 1896 and 1933, a 13-metre (42-foot) wave overwhelmed the 6.5-metre (21-foot) sea wall and swept away more than 80 per cent of the houses, leaving just a single two-century-old pine tree miraculously standing. Nearly 1,900 residents drowned. In Minamisanriku, in Miyagi, the harbour walls proved equally ineffective, and the tsunami washed over four-storey buildings. About 1,200 of the town's population of 17,000 perished, including all but ten of the 130 staff working at the town hall; they were lucky to be thrown by the water towards one end of the roof, where they managed to hang onto a steel pole.

Overall, almost 19,000 people died in the Tohoku earthquake and tsunami, 92 per cent of them by drowning, the rest from building collapse; more than two-thirds were over sixty years old and experienced difficulty in evacuating promptly. A 500-kilometre (320-mile) stretch of coastline was affected, including the total destruction of 129,000 homes (in many cases washed away), the partial destruction of 255,000 homes and the damaging of a further 697,000 homes. About 500,000 people were displaced into temporary shelters, while damage to more than 300 hospitals and other medical institutions forced them to close. The total economic loss is estimated at 16.9 trillion yen ($200 billion), which is about 20 per cent of the annual national budget of Japan. So massive was the challenge of clear-up and reconstruction that when, nine months on, the Dalai Lama paid a visit to Sendai – one of the worst affected areas – he found himself walking among ruins. According to his biographer, Pico Iyer, who accompanied him:

> Houses sat like empty sockets, their first floors shattered and ravaged by the storm (while their second floors sat untouched); telephone poles stood at 45-degree angles, and cars could be seen still floating on the water. His Holiness's car passed crumpled gas stations, houses that were just gaping holes,

huge boats keeled over in the sea. What was once clearly a busy neighbourhood was now a ghost town, rows upon rows of houses buckled over and crumbling, cars piled up in mountains of scrap metal . . .

Then, in a long procession of black-robed monks, to the sound of solemn chanting, His Holiness walked in from the road and slowly up the path towards the local temple, Saikoji, past wreckage on every side, more gravestones crushed or tilted over, greeting a group of kindergarten children, all in blue uniforms, who had been at school the day of the calamity and so survived. Around him trees were torn into stumps and a line of small stone Jizos (the Japanese god of children) sat with red bibs around their necks, protecting the living and the dead.[9]

But it was, of course, the destruction wreaked by the tsunami at the Fukushima Daiichi nuclear power plant that lifted the Tohoku quake into the category of earthquakes that changed the history of the world. The name Fukushima – which ironically means 'Blessed Island' in Japanese – is now globally synonymous with nuclear disaster, along with Hiroshima and Nagasaki. As the Japanese prime minister, Naoto Kan, announced on 13 March 2011, without exaggeration, 'The earthquake, tsunami and the nuclear incident have been the biggest crisis Japan has encountered in the 65 years since the end of World War Two.'[10]

The earthquake at 2.46 p.m. on 11 March, though enormous, was not the chief problem. The power plant was badly shaken, experiencing an acceleration in an east–west direction that exceeded its maximum design basis by 20 per cent; but it withstood this fairly well, because it was built on solid rock. The nuclear reactions in its three reactors then in operation (out of the six in the plant) immediately shut down, triggered by automatic sensors of seismic waves, exactly as planned by engineers. However, alarms in the control room indicated that the plant had now lost its electrical connection to the power grid. To compensate, emergency diesel generators automatically fired up and restored the power supply required to keep the nuclear fuel from overheating.

The chief problem was the tsunami. The plant, with reactors designed in the United States, was first commissioned in 1971, with the example of the

1960 Chile tsunami in mind – like the wall of the ill-fated town of Taro. The JMA's initial tsunami warning at 2.49, indicating a wave height of 3 metres (10 feet) on the Fukushima coast, did not alarm the plant's operators, the Tokyo Electric Power Company (TEPCO), since the sea wall around the plant was 10 metres (33 feet) high. But when the warning was upgraded to 6 metres (20 feet) at 3.14, plant superintendent Masao Yoshida began to worry that there might be damage to emergency seawater pump facilities on the shore, which were required to remove residual heat from the reactors and support equipment, including the water-cooled emergency diesel generators. However, Yoshida anticipated that if these generators failed he could compensate with other available equipment. What he could not anticipate was the full disaster shortly to unfold.

At 3.27, the first wave struck the wall. With a height of 4 metres (13 feet), it was comfortably deflected. At 3.30, as we know, the tsunami warning was again upgraded to a wave more than 10 metres (33 feet) in height in Fukushima. This estimate turned out to be ominously accurate.

What happened next, at 3.35, is described with chilling matter-of-factness in *Fukushima: The Story of a Nuclear Disaster* by two nuclear experts, David Lochbaum and Edwin Lyman, of the Union of Concerned Scientists, based in the US:

> A second wave struck. This one towered about 50 feet, far higher than anyone had planned for. It destroyed the seawater pumps Yoshida had worried about and smashed through the large shuttered doors of the oceanfront turbine buildings, drowning power panels that distributed electricity to pumps, valves and other equipment. It surged into the buildings' basements, where most of the emergency backup generators were housed. (Two workers would later be discovered drowned in one of those basements.) Although some diesel generators stood on higher levels and were not flooded, the wave rendered them unusable by damaging electrical distribution systems. All AC power to [reactor] Units 1 through 5 had been lost. In nuclear parlance, it was a station blackout.

As the wave surged back into the ocean, carrying with it cars, valuable equipment and other debris from the massively damaged plant – observed by the plant's employees from a hillside evacuation point behind the reactor complex – the situation turned desperate. In the words of the experts:

> Japanese regulators, like their counterparts around the world, had known for decades that a station blackout was one of the most serious events that could occur at a nuclear plant. If AC power were not restored, the plant's backup batteries would eventually become exhausted. Without any power to run the pumps and valves needed to provide a steady flow of cooling water, the radioactive fuel would overheat, the remaining water would boil away, and the core would proceed inexorably toward a meltdown . . . Coping with a station blackout is essentially a race against time to restore AC power before the batteries run down.[11]

The AC power supply could not be restored. The first reactor meltdown occurred on 12 March, leading to explosions of hydrogen in three reactors, massive damage to the structures and a giant cloud billowing from Unit 1. All of this was broadcast to Tokyo in real time by Japanese television cameras trained on the plant from a distance, and watched with horror at the headquarters of TEPCO, in the office of the Japanese prime minister and by governments, scientists and citizens of other countries – even as those closest to the action, the residents of the area around the plant, were unaware of what was happening because of the loss of their cell phone and television service. In Washington DC, Lyman, while testifying on the Japanese accident before a US congressional subcommittee on 16 March, warned that: 'We have plants that are just as old. We have had a station blackout. We have a regulatory system that is not clearly superior to that of the Japanese. We have had extreme weather events that exceeded our expectations and defeated our emergency planning measure[s], [such as] Hurricane Katrina.'[12]

The release of radioactive materials put the area within 20 kilometres (12 miles) of the nuclear power plant off limits for local residents, especially in the adjacent town of Okuma; they were advised by the Japanese government on the evening of 12 March to evacuate their homes. On 25 March, those within a radius of 20–30 kilometres (12–19 miles) were also urged to leave. On 12 April, a month after the first meltdown, following further releases of radiation from the plant, the government was forced to raise the international rating of the accident to the highest level, 7: the same level as the Chernobyl nuclear accident in the Soviet Union in 1986. In late April, it ordered the evacuation of an entire village, Iitate – considered one of the most beautiful in Japan – which was outside the 30-kilometre (19-mile) exclusion zone but was threatened by winds blowing from the ruined nuclear plant. In secret, so as not to cause mass panic, the government even considered the evacuation of the capital, Tokyo, in the frantic days after the meltdown when TEPCO argued with the government for complete evacuation of the plant. Yet only in June did the government officially admit that meltdown had occurred at all three reactors, even though it had known this since 12 March, 'based on the off-site detection of tellurium-132, a fission product that could only have come from a melted core', note Lochbaum and Lyman.[13] And only in November, eight months after the accident, did TEPCO at last begin to reveal to the government that the accident was far worse than it had so far acknowledged. Four years later, it was estimated that decommissioning the highly radioactive Fukushima Daiichi plant would take thirty to forty years – at a cost, according to TEPCO, of at least $8 billion.

Could this catastrophe have been foreseen? Could it even have been prevented? The short answer has to be yes to both questions. However, prevention would have required a radical change in policy, both in Japan's nuclear industry and in its government regulation, which was apparently impossible without the spur of the catastrophe.

Until a few years before the 2011 event, Japanese seismologists encouraged corporate and government complacency by arguing against the probability of earthquakes of magnitude greater than 8, 'on the incorrect assumption that different segments of the [Japan Trench] would not break simultaneously, so the largest earthquake on any [segment] would have magnitude 8', notes seismologist Seth Stein.[14] Ocean-floor pressure gauges might have corrected this erroneous view, but at this time few such gauges had been installed. (Many more would be installed after 2011.) Then came the magnitude-9.1 Sumatra–Andaman earthquake in 2004. Some seismologists began to change their mind about the subduction processes in the Japan Trench. By 2008, under pressure from new research that indicated the possibility of great earthquakes on the trench further south than the Sanriku coast, TEPCO was forced to consider the possibility of an earthquake and tsunami off the Fukushima coast comparable to the earthquake that produced the 1896 Meiji Sanriku tsunami. The latter had an estimated magnitude of 8.3–8.5. Based on this magnitude, the company's model predicted a wave of just over 10 metres (33 feet) near the nuclear power plant's emergency seawater pumps, which could sweep inland and reach a height of more than 15 metres (50 feet) around the nuclear reactors. But TEPCO's management dismissed these results as unrealistic, on the grounds that undersea faults and earthquakes in this area of the Pacific were not of the type to produce such large tsunamis.

Soon afterwards, Japanese seismologists raised the further possibility of a repeat of the still-larger Jogan earthquake and tsunami of 869, which may have had a magnitude as great as 9. Moreover, they found evidence of geological deposits from the Jogan tsunami well inland, not that far north of Fukushima Daiichi. TEPCO itself then located other, unidentified, tsunami deposits just north of the plant. But these appeared to be inconsistent with the company's model based on the Jogan earthquake. No action was taken at the plant, such as increasing the height of its sea wall. 'Building embankments as tsunami countermeasures may end up sacrificing nearby villages for the sake of protecting nuclear power stations', an internal TEPCO document concluded. 'It may not be socially acceptable.'[15] When, at a meeting of Japan's Nuclear and Industrial Safety Agency (NISA) in 2009, seismologist Yukinobu Okamura asked a TEPCO representative directly about the Jogan earthquake, the representative attempted to dismiss the question, by stating that this was a 'historic' earthquake not relevant to current conditions. Unwilling to be deflected, Okamura enquired again at the next NISA meeting, citing the massive 2004 Sumatra–Andaman earthquake as a warning. The government's Earthquake Research Committee also took up the question of the Jogan tsunami and the idea of a possible repeat. In response, TEPCO and other Japanese power companies brought pressure to bear on the regulators, pleading with the committee to soften its stand. Not until March 2011, just four days before the Tohoku tsunami, did TEPCO get around to reporting new tsunami damage assessments to NISA.

The company's underlying attitude emerges from its long history of fudging the safety record of its reactors, often in cahoots with government regulators, as relentlessly documented in *Fukushima: The Story of a Nuclear Disaster*. In 2000, for example, a nuclear inspector discovered a crack in a reactor's steam dryer at Fukushima Daiichi. The company requested him to cover up the evidence. When the inspector eventually informed government regulators of the fault, he was fired, and the regulators ordered TEPCO to deal with the problem on its own. In the meantime, Fukushima Daiichi continued in service. Then again, in 2002, the company admitted that for years it had been covering up evidence of cracks in reactor core shrouds at all three of its power plants. Government regulators responded with a public statement that safety was not threatened, which was based on assurances from TEPCO. This time, the company's chairman and president resigned. Nevertheless, both men were retained by TEPCO as advisers. In 2007, an earthquake damaged the company's Kashiwazaki-Kariwa plant on the west coast of Japan: the world's largest nuclear power plant. Following an inspection by the International Atomic Energy Agency, TEPCO had to take seven reactors out of service and improve their seismic resistance. The consequent reduction in its power generating capacity caused the company to make a loss for the first time in three decades. But then a new president, who had spent four decades with TEPCO, returned the company to profit in 2010 with cost-cutting measures. 'His secret?' ask Lochbaum and Lyman ironically. 'Reducing the frequency of inspections.'[16] Even after the Fukushima Daiichi disaster in 2011, astoundingly, TEPCO submitted plans to the government for two more reactors at the plant a mere twenty days after the accident – 'an act of hubris that was not lost on the general public', notes political scientist Richard Samuels in his book, *3.11: Disaster and Change in Japan*. 'The system overestimated safety and underestimated risk because the regulators and the regulated had been in a conspiratorial embrace for decades.' While TEPCO may have been the 'arch-villain', writes Samuels, government regulators – many of whom became well-paid advisers of the electric power industry after retirement – abetted the company's undoubted malpractice.[17]

The above murky history is summarized as follows by Smits: 'long before 11 March 2011, experts in Japan's nuclear power industry knew, or should have known, that a magnitude-9-class earthquake was possible. Indeed, industry officials actively ignored both general and specific warnings.'[18] In an official accident analysis report published in 2012, TEPCO admitted that it did not take preventive action at Fukushima Daiichi after 2008 because it regarded great earthquakes like the Jogan earthquake as hypothetical. From its point of view, perhaps unsurprisingly, the unquantifiable risk to the country did not merit the unquestionable cost to the company. For the government, too, the risk was outweighed by the country's need for nuclear power. Japan is highly industrialized and consumes large amounts of energy; only China, the US, India and Russia, which are much larger countries than Japan, consume more energy. But Japan has few energy resources of its own; it imports 84 per cent of its energy needs. In 2010, the year before the tsunami, nuclear power supplied 29.2 per cent of the nation's electric power. According to the government's Basic Energy Plan, it was projected to supply 53 per cent by 2030, and 60 per cent of all Japan's energy needs by 2100. TEPCO alone had 168 subsidiaries and was Japan's largest issuer of corporate debt by far (7 per cent of the national total). Nuclear power was an industry that was simply 'too big to fail'.

The initial repercussions of the disaster for the industry were enormous in Japan, and significant in the worldwide nuclear power industry; the German government soon announced a phase-out of its nuclear power stations by 2022. Moreover, the tsunami highlighted for the governments of the US and Canada the potentially devastating consequences of an earthquake in the Cascadia subduction zone off the Pacific Northwest, where a magnitude-9.0 earthquake in 1700 produced a massive tsunami that drowned coastal forests in America and damaged the coast of Japan.

Japan's fifty-four working reactors were rapidly closed down for maintenance and review; by May 2012, the country was nuclear free for the first time since 1970. (Although two reactors were restarted later in 2012, they were shut down again in 2013.) Prime Minister Naoto Kan quickly announced that his government would abandon the Basic Energy Plan, suspend all plans to construct new nuclear plants and encourage the development of renewable energy sources. Previously unimaginable public street protests against nuclear power erupted in Tokyo. At shareholder meetings of the utility companies, anti-nuclear resolutions were introduced and seriously discussed. At the TEPCO meeting, under the eyes of riot police, angry hecklers harangued a profusely apologetic management, in one case screaming: 'Go jump in a reactor and die!'[19] An opinion poll conducted by NHK (Japanese Broadcasting Corporation) in June 2011 found that two-thirds of those surveyed thought that nuclear power should be phased out and abandoned. On the first anniversary of the disaster, more than half of those surveyed opposed the restart of the plants and 80 per cent stated that they did not trust government safeguards.

And yet, even in 2011, TEPCO's sclerotic management succeeded in reappointing its board members and chairman, despite the protests of shareholders (though in 2012 TEPCO had to be taken into state receivership). Prime Minister Kan was 'the only principal' in the entire drama who changed his preference from pro-nuclear to anti-nuclear, notes Samuels.[20] By September 2011, Kan was out of office. Since then, the Japanese government has not followed his lead. Kan's successor as prime minister, Yoshihiko Noda, stated frankly in 2012: 'Japanese society cannot survive if we stop all nuclear reactors or keep them halted.'[21] In 2014, the cabinet of the pro-nuclear Shinzo Abe announced a new energy plan in which nuclear power was described as the country's most important power source, projected to account for up to 22 per cent of Japan's electricity by 2030. Although Japanese courts continued to block the restart of most reactors on safety grounds, they granted permission to restart two of them at the Sendai nuclear power plant; one of these was switched on in August 2015. In the longer term, it seems unlikely that Japan – given its demanding energy needs – will want to wean itself off nuclear power, although industry and government will undoubtedly have to upgrade the design standards of nuclear power plants to take account of their risk from great earthquakes and their subsequent tsunamis.

The vacillation over the future of nuclear power is a microcosm of the overall aftermath of the Tohoku earthquake and tsunami. This has become a rerun of events in the mid-1920s following the Great Kanto earthquake, albeit in a more muted form, in which plans for reconstruction in Tohoku and change in Tokyo 'foundered on existing political, administrative and social schisms', exacerbated by a period of weak central government, writes Samuels.[22] To oversimplify a little, three basic political attitudes to the crisis – somewhat reminiscent of the worldwide debate over climate change – have expressed themselves in Japan since 2011. The first attitude maintains that Japan should 'put itself in gear' and move in a new direction, away from its dependence on nuclear power and on the United States. The second takes a more conservative stance, that the disaster was so improbable an event as to be unrepeatable – a proverbial 'black swan' – hence Japan should 'stay the course', maintaining business as usual, if more efficiently conducted. The third argues that Japan must turn 'back to the future', by restoring its core values and its essential identity, which were lost sight of in the rush to modernity since the 19th century and to globalization in the late 20th century.[23] Thus far, elements of all three attitudes have made some public headway: for example, in the Kan government's move to phase out all nuclear power; in the persistence of the central government nexus between business and politics, epitomized by the nuclear power industry; and in the emergence of stronger local government in tsunami-affected Tohoku and some other provinces, as well as the growth of a national volunteering movement from its beginnings in 1995 after an earthquake in the city of Kobe.

There have been significant differences from the 1920s, too. First, as in 1923, there was increased public admiration for the Japanese military, known as the Self-Defence Forces since their establishment in 1954, as a result of their effectiveness in rescue and reconstruction. (In 2011, for the first time, the military were given unrestricted access to nuclear power plants.) But this approval did not raise serious concerns about the militarization of Japanese society, despite the fact that the use of Japanese soldiers to implement martial law in Tokyo in 1923 contributed to the rise of the army to political supremacy in the 1930s. For in 2011 it was obvious to everyone that the attitude of Japanese society to war had changed irrevocably between 1923 and 2011. Second, the behaviour of the Japanese, both in Tohoku and in the rest of the nation, was exceptionally orderly. There was no repeat of the rumour-mongering in 1923 (that led to vigilantism and the murder of Koreans in Tokyo and Yokohama), probably because of the round-the-clock coverage of the disaster by television, not to mention the extensive use of the internet and social media to spread up-to-date information about events and individuals. Thirdly, there was some growth in environmental awareness, symbolized by the launch of Japan's first green political party, committed to the abolition of nuclear power.

The long-term effects of the Great Kanto earthquake are still being debated by historians. However, if the history of the 1923 disaster is any guide, then judging from its political aftermath in the 1920s, it seems that the second of the above attitudes – 'staying the course' – is more likely to predominate in contemporary Japan than the first or the third attitude. As one mainstream Japanese politician, Representative Junya Ogawa, bluntly observed in November 2011: 'Only 20,000 people died in Tohoku, but 30,000 Japanese commit suicide each year.'[24] The Tohoku earthquake and tsunami had not brought Japan to 'a tipping point', he said, unlike the trauma experienced by the country at the end of the Second World War.

CONCLUSION

EARTHQUAKES, NATIONS AND CIVILIZATION

The Trans-Alaska pipeline as it crosses the Denali fault. The pipeline lies on skids, which prevent it from fracturing during Alaskan earthquakes.

Aftershock, a blockbuster film made in China in 2010, opens with a gripping scene from director Feng Xiaogang. The hoot of a railway engine's horn is followed by the sight of a goods train trundling away from the viewer on scruffy railway tracks through a nondescript industrial town. All is quotidian. But then unexpectedly, out of nowhere, the screen swiftly fills with large brown insects and the soundtrack buzzes with the rustle of their beating wings. The ominous swarm appears to pursue the train, almost as if the insects are about to attack it. The crowd of townsfolk seen waiting for the train to pass over the level-crossing – holding their bicycles, many of them wearing Mao caps – are as puzzled as the viewer: what force lies behind this frantic army? 'A storm must be coming,' one ordinary worker tells his excited young son and daughter. An unobtrusive subtitle at the bottom of the screen reads: 'JULY 27, 1976, TANGSHAN CITY'.[1]

The scene is historically true, allowing for some artistic licence. Shortly before the earthquake that destroyed Tangshan on 28 July, there were reports of mysterious dragonfly swarms not in Tangshan itself but in the surrounding region. On 25 July, in the harbour of the port near Tianjin, the insects settled in a dense layer on the windows, masts, lights and sides of a tanker and would not be shooed away. On 27 July, members of one commune spotted a swarm flying north in a perfect formation about 30 metres (100 feet) square. At another commune, a swarm was observed with a breadth of more than 100 metres (330 feet), which took no less than fifteen minutes to pass. 'The buzzing and commotion they made were simply stupefying', writes Qian Gang in *The Great China Earthquake*.[2] Of course, no one had any definite idea about what these swarms, along with some other strange animal behaviour, might portend.

The stunning devastation and even more stunning fatalities of Tangshan's night-time earthquake are skilfully and brutally evoked in the following scenes. However, the rest of *Aftershock* is often melodramatic and less historically true. After the crushing to death of the above-mentioned worker, his narrowly surviving wife locates their two children trapped and severely wounded beneath a single large piece of rubble, and is forced by other survivors to choose between their rescuing either her son or her daughter. In agonized despair, the mother at last chooses her son. But unknown to her, her daughter somehow survives, too, and is cared for by the People's Liberation Army as an orphan. Soon after, Mao Zedong dies in distant Beijing, dutifully mourned by the people of Tangshan. Over the next decade or two, the mother brings up her son as a partially disabled boy who goes on to do well in business, while her daughter is adopted by a childless PLA couple, studies medicine, eventually marries a foreigner and emigrates to Canada. Psychologically traumatized by her abandonment during the earthquake, she makes no attempt to contact her grieving mother, who continues to live in Tangshan, despite its awful memories, and does not remarry.

Meanwhile, the town is rebuilt and, along with the rest of urban China, becomes vastly richer during the economic boom of the 1980s and after. The dusty industrial town of 1976 has been transformed into a city of massive department stores and multi-storey buildings when brother, sister and mother are finally reunited in Tangshan three decades on from the earthquake. This happens as a result of a chance meeting of the siblings after the next major earthquake in China, in Sichuan in 2008, where both of them have come to serve as volunteers in the rescue of the trapped and wounded – working beside international teams. (International help for Tangshan was rejected by a paranoid Mao in 1976, but promptly accepted for Sichuan by the Chinese government in 2008, including rescue teams from Japan and South Korea.) After the daughter forgives her mother, and then comes to appreciate her mother's self-sacrifice while standing at the graveside of her father, the film ends in front of Tangshan's newly built memorial wall for the victims of the earthquake, which was completed just in time for the release of the film.

Thus, *Aftershock* turns the Tangshan earthquake from an unalloyed natural and human catastrophe – the most deadly natural disaster of the 20th century in China – into both a catalyst of economic development and a nationally strengthening event: all made possible by the continuing leadership of the Communist Party (although this is implied, rather than propagandized). The film was enormously successful in China, perhaps unsurprisingly given its excellent special effects, its human drama and underlying patriotic message. However, the truth of its portrayal of the great earthquake and aftermath was sufficient to carry considerable conviction even with an international audience.

'Creative destruction' – a phrase first used in 1942 by economist Joseph Schumpeter in regard to the power of capitalism but here intended to carry a broader meaning – may be said to be part of the aftermath of almost every great earthquake discussed in this book. Creative destruction takes various forms. It can be economic, as in San Francisco after 1906, Tangshan after 1976 and Gujarat after 2001. It can be political, as in Caracas/Venezuela after 1812, Aceh/Indonesia after 2004 and possibly Tohoku/Japan after 2011. It can be cultural, as in Europe, after the Lisbon earthquake in 1755, which provoked Voltaire and Enlightenment philosophers, and after the Naples earthquake in 1857, which kick-started seismology. One might even argue that Akira Kurosawa's formative adolescent experience of the Great Kanto earthquake in Tokyo in 1923 – witnessing scenes of violent mayhem and mass annihilation – was vital to the creation of his classic films about the extremes of human behaviour and emotion. Moreover, creative destruction succeeds to varying degrees. It was perhaps most successful in San Francisco after 1906, when new construction, business and innovation all flourished. It was less successful in Tokyo after 1923, when the opportunity to redesign the city was largely lost, a financial panic ensued and eventually the army took over the government. And in Lisbon after 1755, it was still less successful; the earthquake led directly to the imposition of a dictatorship, and in due course the economic decline of Portugal and its empire. Yet even in Lisbon, the ruined city was redesigned to a new plan that is much admired to this day. 'Disaster spurs reinvestment and creative destruction as long as the source of urban economic strength remains fundamentally unaffected. Capitalism, in this sense, outflanks catastrophe', note two urban planners, Lawrence Vale and Thomas Campanella, in their conclusion to *The Resilient City*, a study of how modern cities recover from disaster.[3] The urban economic strength of early 20th-century California was on the rise, whereas that of mid-18th-century Portugal was on the decline.

At the furthest extreme of this spectrum of success was the magnitude-7.0 earthquake that struck Port-au-Prince, the capital of Haiti, one of the poorest nations in the world, in 2010. The mortality rate – over 300,000, according to the Haitian government – was horrendous. The estimated total direct loss from collapsed buildings was $8 billion, of which only about 2.5 per cent was insured, notes Anselm Smolka of the Munich Reinsurance Company. 'The country was flooded with donor capital, but in the end any efforts to organise a concerted action for immediate disaster relief, for a well-organised reconstruction process, and to restore a functioning public administration failed', writes Smolka. 'Notwithstanding partial successes and the admirable engagement of many individuals and aid organisations, disaster recovery after the Haiti earthquake must be seen as a missed opportunity. This is all the more embarrassing as the chance of further earthquakes happening even closer to the city is high' – according to seismologists.[4]

In fact, the Haiti earthquake had been foreseen in a seismological paper published in 2008. After five years of collecting Global Positioning System (GPS) data about increasing strain on the little-known Enriquillo–Plantain Garden fault – which runs along the southern side of the island of Hispaniola, where Haiti and the Dominican Republic are located – seismologists concluded at a conference in the Dominican Republic that a magnitude-7.2 earthquake was probable if the fault were to release all of its stored energy. Moreover, the leader of the group, Eric Calais, repeated this conclusion at private meetings with the Haitian government and the Haitian Bureau of Mines and Energy during 2008, while acknowledging that the timing of the earthquake was impossible to forecast, even approximately. Inevitably, the Haitian government, with numerous other priorities, did nothing to plan against a future earthquake disaster.

That said, the proper response of governments anywhere in the world to such earthquake 'predictions' is deeply problematic – even in the most advanced nations. In 1994, a magnitude-6.7 earthquake shook Northridge, a neighbourhood of Los Angeles in the San Fernando Valley, causing damage estimated at $20 billion – that is, the most costly earthquake disaster in the history of the United States. Very shortly after the earthquake occurred, 'when the southern California earth science community was still abuzz with the frantic adrenaline rush that accompanies large and destructive local earthquakes', notes Susan Hough in *Earthshaking Science*, someone asked 'an eminent seismologist' if anyone had predicted the earthquake. 'Not yet,' he replied ironically.[5]

No doubt he had in mind the Parkfield prediction, which was much discussed in the early 1990s. Parkfield, a tiny community situated on the San Andreas fault half-way between Los Angeles and San Francisco, was for some years from the late 1980s the self-styled 'Earthquake Capital of the World', where the earth, allegedly, 'Moves for You'. It boasted the only officially endorsed earthquake prediction in the US. The recurrence interval of Parkfield earthquakes seemed to be about twenty-two years: moderate earthquakes were reported there in 1857 and 1881 and were scientifically recorded in 1901, 1922, 1934 and 1966 (the last one serendipitously recorded by seismographs that had been deployed to monitor an underground nuclear explosion in Alaska). In 1985, the United States Geological Survey announced that there was a 95 per cent probability of a magnitude-6 earthquake occurring at Parkfield before the end of 1992.

Unfortunately, seven years and $18 million later, the Parkfield area had proved notable chiefly for its seismic *in*activity. The largest quake was one of magnitude 4.5 in October 1992. Promptly, the USGS issued a warning that the expected magnitude-6 quake might follow within seventy-two hours. The California Office of Emergency Services set up a mobile operations centre outside the Parkfield Café. In towns nearby, fire engines stood ready, and residents laid in extra supplies of water. Helicopters from several television stations hovered overhead, and reporters from dozens of newspapers arrived on the scene. But there was no sign of the earthquake. It finally arrived in 2004, thirty-eight years after the 1966 one. Except for the fault rupturing from southeast to northwest, rather than the other way around, as in 1922, 1934 and 1966, the extent of the rupture and the magnitude (6.0) were as expected. But as an example of a prediction, Parkfield's has to be regarded as a very qualified success, at best.

Earthquake prediction is a seductive mirage – forever beckoning but always out of reach. Among professionals, periods of optimism about prediction have alternated with periods of pessimism since the start of seismology. Reputable seismologists and geologists have quite often been tempted into making predictions. In Japan, as we know, back in 1905 seismologist Akitsune Imamura correctly predicted the epicentre of the Great Kanto earthquake in 1923 (under Sagami Bay) and got the timing right, within a fifty-year window. But there was no reliable theory underlying his prediction. In 1911, the pioneering John Milne remarked in the journal *Nature*: 'The popularity of the seismologist would be enhanced if, like the astronomer, he had the power to predict . . . Astronomers have received the support of nations since the days of astrology, while seismology is in its childhood seeking for more extended recognition.'[6]

In 1980, when optimism was running high, a geologist at the US Bureau of Mines predicted that a giant earthquake would strike Peru in June 1981. Brian Brady had been studying quake-like rock bursts in mines. Rock bursts occur when mining activity reduces the confining pressure on neighbouring rock. Brady's examination of rocks as they fractured in the laboratory had convinced him of the existence of a 'clock' in the fracture process: once started, it would inexorably run on and produce an earthquake, its ticks being bursts of moderate foreshocks. Given the requisite historical and current seismicity data, Brady said he could tell exactly when the quake would occur. Fellow geophysicists refused to accept this theory that the small-scale (microscopic) process of rock fracture and the large-scale (macroscopic) mechanism of earthquakes were basically the same – 'scale invariant' in scientific language – since geological faults, unlike mine walls, always remain under enormous confining pressure. They said so formally through the National Earthquake Evaluation Prediction Council in early 1981 at the end of what amounted to a 'trial' of Brady before his peers. The prediction had been news before; now it was headline news. But Brady refused to withdraw it, and so it became thoroughly entangled in Peruvian–US politics. The Peruvian president, government and scientific community were taking it seriously, while in Washington DC, and in the offices of the Geological Survey and the Bureau of Mines, several different groups were vying to use the prediction for their own ends. The people of Peru, 66,000 of whom had died in an earthquake as recently as 1970, became more and more jittery as 28 June 1981 approached. The capital, Lima, became eerily quiet: many people, both rich and poor, left town for the weekend. Nothing happened.

On the other hand, in 2009, following frequent tremors in central Italy's quake-prone Abruzzo region, Italian scientists predicted that a major earthquake was unlikely. A week later, a magnitude-6.3 quake struck the regional capital, L'Aquila, causing 309 deaths and wrecking much of the city. Some of those killed by building collapse had stayed indoors after listening to the public reassurances of scientists and government officials. These individuals were now charged with negligence by the city's authorities in what became a notorious court case, where science itself appeared to be on trial.

The fact that earthquake prediction as a science has had such a chequered history is hardly to be wondered at. 'One may compare it to the situation of a man who is bending a board across his knee and attempts to determine in advance just where and when the cracks will appear', wrote Charles Richter in 1958. 'All claims to predict the future have a hold on the imagination; it is not surprising that even qualified seismologists have been led astray by the will-o'-the-wisp of prediction.'[7] About the predictors themselves, Richter wrote bluntly in some unpublished notes in 1976: 'What ails them is exaggerated ego plus imperfect or ineffective education, so that they have not absorbed one of the fundamental rules of science – self-criticism. Their wish for attention distorts their perception of facts, and sometimes leads them on into actual lying.'[8] (Here, however, Richter was probably referring more to amateur predictors than to his seismologist colleagues.)

Scientists' hopes for long-term prediction are pinned mainly on a cyclical concept arising from the 'elastic rebound' model: fault stress is thought to build at a constant rate and then dissipate abruptly in regularly occurring ruptures. In the short term, prediction must depend on precursors and, by extension, the instrumentation, personnel and social organization necessary to observe and measure them. Possible precursors include foreshocks, changes in ground strain, tilt, elevation and resistivity, alterations in local magnetic and gravitational fields, shifting of ground water levels, the emission of radon gas, deep sounds, flashes of light and the peculiar behaviour of animals. Some of these precursors may appear months, or even years, ahead of a large earthquake; others only in the days and hours before it strikes.

Foreshocks are the most useful precursors. But as Roger Musson warns, 'What marks them out as foreshocks is the fact that a big main shock happens afterwards. Until the main shock occurs, they seem just like any other small earthquake.'[9] Still more awkwardly for predictors, foreshocks frequently do not happen, at least not in the period immediately before the quake. There was no foreshock in Tokyo in 1923, none in Tangshan in 1976, practically none in Gujarat in 2001, and none in a typical major Californian quake such as the one at San Fernando in 1971 (magnitude 6.5). A later review of micro-earthquakes in the area during the thirty months leading up to the San Fernando quake showed, however, that a drop in the speed of P (primary) seismic waves by 10–15 per cent had taken place – followed by a return to normal speed just before the quake. A rather similar phenomenon had been observed by Soviet seismologists after detailed monitoring of small and large earthquakes in Tajikistan during the 1950s–60s: they found that the ratio of P to S (secondary) seismic wave speed dropped for a variable period and then suddenly returned to normal just before a major quake. US monitoring, spurred on by the Soviet results, seemed to confirm this general picture, and for a while optimism about earthquake prediction soared. The magazine *Scientific American*, introducing an article on the subject in 1975, went so far as to declare: 'Recent technical advances have brought this long-sought goal within reach. With adequate funding several countries, including the US, could achieve reliable long-term and short-term forecasts in a decade.'[10] However, subsequent extensive measurement of seismicity at the San Andreas fault revealed no such general predictable behaviour by P waves. If this method does prove useful in prediction, it will work (like so many other methods in seismology) only locally, within particular geological conditions that have been extensively studied over a sufficiently long period.

This is certainly true of attempts to understand uplift and subsidence of the ground. On 16 June 1964, a major earthquake struck the coastal town of Niigata in western Japan, with its epicentre just off the coast at Awashima Island. There was a sudden subsidence of the coastline by 15–20 centimetres (6–8 inches). This in itself was unremarkable, but when plotted on a graph of land elevation in relation to mean sea level since 1898, the sudden slump was shown to have followed a gradual rise in the land opposite Awashima Island at the rate of nearly 2 millimetres (a sixteenth of an inch) a year. The fact was of course noticed only after the earthquake. By keeping a constant watch on the elevation of likely trouble-spots, with the help of laser-ranging devices and the satellites of the GPS, uplift may eventually turn into a useful indicator of a coming earthquake.

But even if uplift can be successfully quantified in advance of a disaster, there remains the tricky question of interpretation. The most celebrated example is the 'Palmdale Bulge', the uplift of an area of southern California centred on Palmdale, some 72 kilometres (45 miles) north of Los Angeles, and extending 160 kilometres (100 miles) along the San Andreas fault. The uplift, measured from the 1960s (that is, before the advent of the GPS), was said to amount to a striking 35 centimetres (14 inches), though subsequent studies suggested this figure had arisen from survey errors, leading to heated arguments in the 1970s about whether or not the bulge really existed. If it did, what did it signify – perhaps the imminence of an earthquake in the area, which lies on the southern section of the fault, the 300 kilometres (190 miles) that have not slipped since 1857? The final verdict seems to be that there is solid evidence of some uplift in the Palmdale area, and that it is a consequence of the Kern County earthquake in 1952, notes Hough. 'But as a harbinger of doom it [has] clearly not lived up to expectations.'[11]

Abnormal animal behaviour before some earthquakes has yet to be proved by a controlled experiment. Research has shown no correlation between cockroach activity and impending earthquakes, and no response from cows wearing earthquake sensors – even to the earthquake itself! There is a correlation between the number of lost pet advertisements and major storms (during which animals may run away), but no correlation between the number of such ads and major earthquakes. The design of animal-oriented earthquake experiments encounters some obvious difficulties, compounded by the fact that seismologists are instinctively sceptical about anecdotal reports and prefer scarce research funding to be spent on more potentially productive investigations.

Nonetheless, there is ample evidence that animals may perceive an earthquake coming, as in Haicheng in 1975 and Tangshan in 1976. Reports exist from all over the world, and date back to the earliest times, as documented by scientist Helmut Tributsch in his book, *When the Snakes Awake*.

In antiquity, the historian Plutarch mentions a rabbit having had a premonition of the earthquake in Sparta in about 464 BC, and Pliny the Elder describes something similar in his *Natural History*. In 1755, Immanuel Kant noted of the Lisbon earthquake:

> The cause of earthquakes seems to spread its effect even into the surrounding air. An hour earlier, before the earth is being shaken, one may perceive a red sky and other signs of an altered composition of the air. Animals are taken with fright shortly before it. Birds flee into houses, rats and mice crawl out of their holes.[12]

Before the 1923 earthquake in Tokyo, catfish (the mischievous *namazu* of tradition) were seen to jump agitatedly in ponds, and could be caught in bucketfuls. In China, the appearance of panicky rats is an officially designated seismic precursor. In May 1974, according to a scientific report, this precursor saved the lives of a family in Yunnan province. A housewife had found rats running about her house since 5 May. On the night of 10 May, they were so noisy she got up to hit them. Then she suddenly recalled a visit to an exhibition on earthquakes and evacuated the house. The following morning, a magnitude-7.1 earthquake occurred, and the house collapsed.

Explanations of these and thousands of other reported instances may involve better-than-human sensitivity to vibrations and sounds, electrical and magnetic fields and the odour of leaking gases. An electrical explanation – perhaps animals' exposure to clouds of charged particles emitted by the ground – seems the most likely. If the mechanism were ever to be established, and an analogous, affordable detection instrument designed, then animals would no longer be useful for prediction – just as canaries are no longer needed to detect gases in mineshafts.

According to a trio of Greek scientists – two solid-state physicists and an electronics engineer – studies of the electrical resistivity of the ground may throw some light on phenomena like the Palmdale uplift, and also provide a method of earthquake prediction. The VAN method, named after its inventors Panayotis Varotsos, Caesar Alexopoulos and Kostas Nomikos, is based on a fact first reported by Milne in an 1898 paper before Britain's Royal Society: ahead of a powerful earthquake, the natural electrical currents that circulate and fluctuate in the ground (telluric currents) are disturbed, and hence the resistivity of the ground alters. Efforts to detect this so-called Seismic Electrical Signal (SES) and use it for prediction failed in many countries, and the approach was abandoned. But in the 1980s, the Greek scientists took it up and claimed substantial results: during 1988 and 1989, they predicted the location and magnitude of seventeen earthquakes around Greece with some success.

The VAN method is highly controversial, though it has some scientific supporters all over the world. One difficulty, common to seismology as a whole, is that each area of the world presents unique problems of data interpretation. Another is the long time (decades) required to calibrate a network of VAN stations in areas where, unlike Greece, tremors are uncommon. A third difficulty is the lack of a satisfactory explanation of the SES. But the most serious difficulty is that the SES, according to the VAN hypothesis, is not universally present during earthquakes; it is detectable only at certain 'sensitive sites'. As a consequence, 'no amount of negative evidence – the absence of the SES prior to large earthquakes – can ever disprove the hypothesis,' writes an unconvinced Hough, 'because any and all negative results can be dismissed as having been recorded at insensitive sites.'[13]

As for long-term earthquake forecasting, it is far less successful than long-term weather forecasting – and potentially a great deal more hazardous. In Musson's graphic comparison: 'We can track every weather system, every cloud. Yet we are still only so-so at predicting the weather. Now imagine that all the clouds, all the weather systems, are miles underground, out of sight. Want to try making a weather forecast now? Well, that's how it is with earthquakes.'[14] Geological processes are so slow that prediction, even on the basis of a century of data, is like trying to predict tomorrow's weather on the basis of one minute's observation. Until the arrival of plate-tectonic theory in the 1960s, about all that could confidently be said was that earthquakes mostly would occur where they had previously occurred. Today's theory focuses this statement a little by suggesting that the longer the period since an earthquake has occurred, the more likely that place is to experience a quake. Many scientists also reckon that the magnitude of the quake increases with increased time of quiescence of a fault.

At Pallett Creek, for instance, 55 kilometres (35 miles) northeast of Los Angeles, geologist Kerry Sieh dug a trench into the San Andreas fault in the 1970s, and revealed well-differentiated strata of silt, sand and peat that appeared to have been disturbed by a series of large earthquakes over the past 1,400 years. Using carbon dating, he established the following dates for these palaeoseismic earth movements, all but the last of them approximate: 545, 665, 860, 965, 1190, 1245, 1470, 1745 and 1857. The greatest interval is 275 years, the smallest 55 years, while the average is 160 years. Will southern California experience its next large earthquake during the next decade (given that 1857 + 160 = 2017), or merely during this century? The recurrence interval is clearly too variable for any meaningful prediction. In northern California, on the Hayward fault near San Francisco, the situation seems slightly clearer. Another geologist, James Lienkaemper, has excavated trenches that show twelve large (magnitude-7) earthquakes during the past 1,650 years – the latest in 1868 – with an average recurrence interval of 140 years over the last five of these quakes. But this is not sufficient evidence to predict the timing of the next big earthquake on the Hayward fault – as the following analysis shows.

In 1979, four geophysicists defined segments of plate boundaries around the Pacific Ocean that had not experienced large earthquakes for thirty years as 'seismic gaps', to which they allotted a high potential for a future large earthquake. But in the decade after 1979, of the thirty-seven earthquakes with magnitude 7 that occurred in the north Pacific, a mere four quakes occurred in these seismic gaps, whereas there were sixteen quakes in zones predicted to have intermediate potential and seventeen in zones predicted to have low potential. Moreover, when only the largest earthquakes (magnitude 7.5 or greater) were considered, just one quake occurred in the high-potential zone, as against three in the intermediate-potential zone and five in the low-potential zone. The fit between prediction and outcome would have been better had the zones been assigned their potentials randomly. 'The apparent failure of the gap model is surprising, given its intuitive good sense', commented Seth Stein in *Nature*. 'It may be that seismicity in some regions is quasiperiodic, whereas in others it clusters.'[15]

The whole field of earthquake prediction is plainly wide open for speculation. Many predictions are made by unscientific and pseudoscientific means, and most of these are ignored. Occasionally, though, for reasons that are not always clear, one of them catches on and causes panic. This happened in Haicheng in 1975, and proved to be prescient. But it also happened in December 1989, when a self-taught climatologist, Iben Browning, predicted that a subtle bulging of the earth caused by the gravitational pull of the sun and the moon – calculated by astronomers to peak on 3 December 1990 – would trigger a catastrophic earthquake in the Mississippi valley, comparable to the earthquakes of 1811–12 around New Madrid, Missouri. In addition to having a PhD (in zoology) and the support of the director of Southeast Missouri State University's Earthquake Information Center, Browning was said – by the *New York Times*, no less – to have forecast the major earthquake in San Francisco on 17 October 1989 'a week in advance'.[16] The *San Francisco Chronicle* stated: 'He missed by just 6 hours hitting the Oct. 17 San Francisco quake on the nose in a forecast published in 1985 and by only 5 minutes in an update a week before the disaster.'[17] As the director of San Francisco's Bay Area Earthquake Preparedness Project later remarked, 'these things have a life of their own.'[18]

Browning's prediction caused frenzy for months in the Midwest. People in Missouri spent $22 million on earthquake insurance. On the predicted day itself, the governor of Missouri and a national media circus descended on New Madrid. The public reaction was not much affected by rational arguments, such as those of a committee of seismologists who examined Browning's October 1989 California earthquake prediction in a video-recording and transcript and found it to be baseless, or the fact that the supportive Missouri seismologist, despite having a PhD in geophysics, was known to believe in psychic phenomena. In a short story about the area published a decade later, 'A Comparative Seismology', a conman posing as a USGS seismologist tells a lonely older woman: 'The New Madrid fault's been dormant for nearly 200 years. Seismic tension building by the day. That's why you don't feel any action – there hasn't been any. But there will be eventually, Miss Silver. You have my scientific guarantee.'[19] She believes him, takes up his offer of escape from the coming catastrophe, and loses a lot of money.

The real scientific community could probably have scotched the Browning prediction at the outset, but it failed to act quickly. This was partly because it did not take the forecast seriously, partly because of its general lack of confidence in the scientific approach to the subject, especially in regard to earthquakes in the Mississippi valley (as opposed to better-monitored California), and partly because of internal politics between federal agencies such as the USGS and university-based seismologists. One of the latter, Stein, refused to be drawn into the media circus. Yet Stein admits in his thought-provoking and balanced book on earthquake hazard in the Midwest, *Disaster Deferred*, that scientists bear some responsibility for the hyperbolic public response in 1989–90: 'Browning's prediction was the spark that set off prepared firewood. The firewood was what federal and state agencies and even some university scientists had told the public about past earthquakes and future hazard. Much of this was wildly exaggerated to the point of being embarrassing.'[20]

Unfortunately for scientists, the elastic rebound model of earthquakes is insufficient, on its own, to explain how earthquakes occur. Brady's rock-burst theory is even less effective, hence the failure of his prediction. Yet, Brady was prescient in this defence of his position in 1981:

> Many within the seismological community are currently infatuated with simple fault models made more complex by the addition of asperities (hard zones along the fault surface) which tend to inhibit free body motion along the fault; an earthquake occurs once the asperities are broken . . . I believe we need to address the fundamental problem of *how* the fault gets there in the first place.[21]

Only now have seismologists begun to face this challenge head on. One of them compared himself with rueful candour to an 18th-century physician, 'who although lacking understanding of disease is compelled to do something and so prescribes bleeding.' The scientific problem of understanding the origin of faults, he remarked, is likely to get worse before it gets better: our expanding knowledge of the earth, derived from the extraordinary sophistication of new instrumentation, has ironically 'served to magnify our lack of understanding'.[22] Perhaps nowhere illustrates the truth of this remark better than the earthquakes of the American Midwest. What is the origin and nature of the invisible intraplate faulting that produces such earthquakes 'forbidden' by plate-tectonic theory? Thanks to extraordinarily accurate GPS surveys of the New Madrid area of Missouri launched in the early 1990s, we now know that the North American plate shows less than 2 millimetres (1.5 inches) of movement per year (as compared with an average of about 36 millimetres, or 1½ inches, per year for the San Andreas fault). In other words, it is virtually stationary. Does this mean that another large earthquake like those of 1811–12 is in the offing – or does it mean the opposite: that the area should be considered seismically inactive? The stakes are high for scientists, government agencies and residents; the debate is intense; but the truth, as ever with earthquake prediction, is likely to remain elusive for many years to come.

While scientists from various disciplines investigate and theorize about earthquakes and earthquake prediction, what can governments, institutions and individuals do to protect themselves from the seismic menace? Nearly half of the world's big cities now lie in areas at risk from earthquakes. For example, the capital of Iran, Teheran, with a population of more than 8 million people, is built on top of fifteen active faults, which are known to have produced more than ten earthquakes of magnitude 7.0 or greater in the historical record. Iran as a whole has experienced fourteen earthquakes of this magnitude during the past century alone (not to mention the magnitude-6.6 quake in 2003 that levelled the city of Bam).

Areas at risk also include cities and countries where the level of seismicity is much lower but earthquakes still spring an occasional unpleasant surprise. Egypt – whether ancient or modern – has not been much known for earthquakes. Yet in 1992, Cairo was struck by a magnitude-5.8 quake with its epicentre just 10 kilometres (6 miles) south of Old Cairo. It killed 545 Egyptians, injured some 6,500 and made 50,000 homeless, as well as completely destroying 350 buildings and severely damaging 9,000 others, including 350 schools, 216 mosques (the upper part of one of the minarets of al-Azhar collapsed) and ancient monuments such as the Great Pyramid at Giza (from which a large block rolled to the ground). Part of the reason for the disproportionate loss of life and structural damage was that Cairo had not been struck since 1847. As a result, the city had no building regulations in place to reduce damage by earthquakes and no contingency plan for the inhabitants in the event of an earthquake. Most of those who perished were the impoverished tenants of poorly constructed apartments or children crushed in the stampede to escape collapsing schoolrooms. Large parts of the city with better-constructed buildings were left untouched.

'Earthquakes don't kill people; buildings do.' This saying has been popular with seismologists since it was used in a lecture by Nicholas Ambraseys in 1968. It is especially pertinent to substandard buildings. 'The assembly of a building, from the pouring of foundations to the final coat of paint, is a process of concealment, a circumstance ideally suited to the omission or dilution of expensive but essential structural components', Ambraseys and fellow seismologist Roger Bilham noted in 2011, on the first anniversary of Haiti's earthquake, in an article entitled 'Corruption Kills'.[23] In Port-au-Prince, the death toll in 2010 was particularly high because many buildings pancaked as a result of their supporting columns being made of substandard concrete or of cinder blocks lacking adequate steel reinforcement. Some 83 per cent of deaths from building collapse in earthquakes during the period 1980–2010 happened in countries that are anomalously corrupt, according to calculations made by Ambraseys and Bilham. 'You can bribe a building inspector – but you can't bribe an earthquake', writes Musson.[24] Whether as a result of corruption or of other constructional weaknesses, around the globe hundreds of millions of lives are now at permanent risk from earthquakes, along with countless billions of dollars worth of property – numbers and amounts that are certain to rise further.

By contrast with Teheran, Cairo and Port-au-Prince – indeed almost all of the cities at seismic risk – in Tokyo, a barrage of scientific instruments is in place to monitor aspects of the many faults that may affect the metropolitan area; and these are connected to a high-tech control centre. Since 1977, an emergency committee of scientists has been at permanent ready to respond to unexpected movements of the crust and advise the Japanese government whether to issue an alert. The government has designated evacuation areas and briefed the population through heavy publicity; every 1 September, the anniversary of the Great Kanto earthquake, there are city-wide earthquake drills. Stringent building regulations have long been in force. Most major Tokyo buildings have been retrofitted against earthquakes and new ones have for many years routinely been constructed to withstand the maximum possible shaking. Tokyo's skyscrapers and tower blocks are supposed to be the safest buildings in the city; indeed, office workers and residents are advised to stay inside them in an earthquake, no longer to rush out and risk being cut by flying glass or killed by one of the shop signs that hang above the streets. This recommendation proved accurate during the Tohoku earthquake in 2011.

The city's population has, however, yet to be tested as it was in 1923, by a great earthquake with its epicentre close to the capital. Despite the annual earthquake drill, most of those living in Tokyo do not give much thought to what might happen, judging from informal questioning of residents around 1990 by journalist Peter Hadfield for his book on the subject, *Sixty Seconds That Will Change the World*. A 27-year-old Japanese fashion designer, who moved to Tokyo to make his career, told Hadfield: 'I don't know how many people will be killed. A million? A billion? I really don't know. It's not the sort of thing I ever talk about with my friends. I know an earthquake is possible, but deep in my heart I can't really believe it.' A 25-year-old businessman working for an international food company was a bit more realistic: 'I'm not so worried about it. I don't talk about it so much with my friends – only sometimes, when we're driving through a tunnel and I might make a joke about it. I don't know how many people will die. It depends on the severity of the earthquake. Maybe 2 million?'[25] According to Hadfield, 'The majority of ordinary Japanese I spoke to during the course of researching this book had the feeling that, although they know and have been told that a major quake is coming, they still find the reality of such an event hard to believe.'[26]

The deaths, damage and slow government response after the unpredicted, magnitude-6.8 earthquake in 1995 in Kobe, Japan's sixth largest city, 500 kilometres (320 miles) to the west of Tokyo, were not reassuring. Not only did 6,400 people die, 55 per cent of the structures built before 1970 collapsed, including the supposedly earthquake-resistant Hanshin expressway, as a result of a maximum horizontal ground acceleration equal to 80 per cent of the acceleration due to gravity: twice that of the ground acceleration estimated for the Great Kanto earthquake. For the first time, many Tokyoites realized the high risk that they themselves were running. Novelist Haruki Murakami's response – a short-story collection with the lower-case title *after the quake* – opens with a married woman in Tokyo glued to a television set for five whole days, 'staring at crumbled banks and hospitals, whole blocks of stores in flames, severed rail lines and expressways' in quake-struck Kobe. This nihilistic media exposure drives the woman to divorce her husband, abandoning him with just a note on the kitchen table saying that 'living with you is like living with a chunk of air'.[27] When her former husband later tries to have sex with a woman, he gives up, because he cannot forget the silent images of 'highways, flames, smoke, piles of rubble, cracks in streets' running through his mind like a slideshow.[28]

Probing Tokyo's physical reality and protection shows how many loopholes remain, and, more seriously, how vastly more vulnerable the modern city is than the Tokyo of 1923. What about the bullet trains, which travel at speeds of 240–300 kilometres (150–90 miles) per hour? Train braking systems triggered by seismographs can give bullet trains some seconds of advance warning of a coming shock wave and bring them screeching to a halt. This is what happened in March 2011, when twenty-seven bullet trains avoided derailment by applying emergency braking nine seconds before the Tohoku earthquake's shaking began, and seventy seconds before the most violent shocks. But will the brakes react fast enough if a train happens to be close to the epicentre in a severe Tokyo earthquake? What about the refinery and chemical complex built on soft reclaimed mud beside Tokyo Bay, the new tower blocks built there against expert advice, and the Hamaoka nuclear power plant, built in the 1970s near the junction of two tectonic plates, 200 kilometres (125 miles) southwest of Tokyo? (This plant was shut down after the Fukushima nuclear disaster.) What about corruption in the construction industry, tellingly portrayed by Kurosawa in his film, *The Bad Sleep Well*, half a century ago, and still doing well, as we know, in the nuclear power industry in 2011? What will happen to electricity cables, gas mains, water mains, telephone and computer lines, communications in general? Nor should the risk of a fire be overlooked in the large traditional areas of Tokyo with narrow lanes and predominantly wooden housing, such as were incinerated in 1923. And who will coordinate the rescue efforts and cut through the bureaucratic rivalries that bedevil the emergency planning?

The habit of building on poor foundations in earthquake country is, needless to say, a common one. Dozens of examples from many civilizations throughout history are detailed in *Apocalypse* by Amos Nur (who narrowly escaped serious injury beneath crashing steel bookcases in his office at Stanford University during the 1989 earthquake near San Francisco by diving under his desk). Hence the half-ruined state of Rome's Colosseum due to an earthquake, probably in 1349: the northern section of the ancient amphitheatre's external wall remains standing, while the southern has collapsed. A seismic study of the Colosseum's foundations, using sound waves to create images of the subsurface structure, revealed in 1995 that the southern half rests on alluvium – accumulated sediment filling the prehistoric bed of a tributary of the River Tiber that is now extinct, whereas the northern, undamaged, half stands on the riverbank – where the ground is older and more stable.

Over the centuries, much has been learnt about earthquake-resistant construction by trial and error. Hence the evolution of the complicated wooden joints that support the roofs of Japanese pagodas. 'The structure of a pagoda, with a central pillar set into a foundation stone and the wooden multistoried structure arranged around it with bracketing, enabled the building to shake independently around the central pillar', notes art historian Gina Barnes.[29] In 1923, the 17th-century Kan'eiji Temple pagoda in Tokyo was one of the few buildings to survive the earthquake intact. Its design apparently inspired the flexible steel-frame lattice for Japan's first skyscraper, the Kasumigaseki building in Tokyo, created by engineer Muto Kiyoshi in 1968. In Turkey and Kashmir, notes Susan Hough, 'people have long recognized creaks and cracks as an effective defence against earthquake damage', because these imperfections help to prevent a building from destructive swaying. 'Traditional architecture in these regions incorporates a patchwork quilt of wood elements and masonry infill, producing buildings that are able to dissipate shaking energy in a million little internal shifts and shimmies.'[30] At the church of Hagia Sophia in Istanbul (the former

Constantinople), the greatest of the Byzantine churches, the architect-engineers of the 6th century AD used a flexible cement to allow the walls of the building to 'give' a little during earthquakes. They added volcanic ash or other silica-rich materials to their mortar of limestone and crushed brick. This reacted with the limestone and water and produced a calcium silicate matrix – similar to that found in modern Portland cement – that can absorb seismic energy.

Modern earthquake-resistant design involves structures with steel frames and reinforced concrete – because steel bars have great strength in tension (but none in compression), whereas concrete has great strength in compression (but none in tension) – and 'shear walls', that is, strong walls that prevent a building from shifting laterally too far. A shear wall was constructed in the Seismological Laboratory of the California Institute of Technology at Pasadena, soon after hospital buildings collapsed in the destructive San Fernando earthquake of 1971, so as to prevent potential future embarrassment to seismologists. A more recent innovation in earthquake engineering, known as 'base isolation', employs rolling rubber or lead bearings between the building and its foundation, in order that much of an earthquake's horizontal ground motion is not transmitted to the building. In Alaska, where the crucial Trans-Alaska Pipeline passes over the Denali fault, sections have been engineered so as to lie on skids. When a magnitude-7.9 earthquake struck there in 2002, this fault moved 7 metres (23 feet), but the pipeline did not break. (Given the cost of the special engineering, however, the average pipeline is allowed to break in an earthquake, and then repaired as soon as possible.)

Designs are tested for their probable performance during an earthquake in one of three ways. In the first place, the likely movement of a building can be calculated from formulae based on the overall size, 'stiffness' and other properties of its structure. Second, the building can be simulated on a computer and subjected to a simulated shaking. Third, a scale model of the building can be made and physically shaken on a so-called 'shake table'. Besides the obvious expense of the last method, it suffers from another limitation, that it may not be scale invariant: a small-scale model may respond differently from a life-size building to the same shaking. Nonetheless, maybe a dozen large shake tables are in use around the world – one of which, at the University of California in San Diego, has a 93-square-metre (1,000-square-foot) steel platform large enough to test a full-scale slice of a seven-storey building, with a maximum load of 2,000 tons.

A building's natural period of oscillation – in which it would sway back and forth if you gave it a push (like a playground swing) – is important in earthquakes. The natural period of a ten-storey building is about one second, and this increases by about one second for every ten additional storeys. Skyscrapers therefore have a longer natural period than low-rise buildings. With a short period of vibration of, say, one-tenth of a second in horizontal ground shaking, an earthquake will make furniture and other objects inside a building rattle, but leave the structure unmoved. With a long period of, say, ten seconds, the whole building will move as one, without swaying significantly. But if the period of the vibration matches the natural period of the building, the two will be in resonance – like a swing pushed at just the right moment of each oscillation so as to make it go higher and higher – and the building will sway. If the shaking persists, there will be a strong likelihood of building collapse.

More crucial, however, to the proneness of a building to collapse is the choice of construction material (and of course the quality of construction). Reinforced concrete buildings generally survive best in earthquakes, timber-frame buildings next best, brick buildings less well than timber-frame buildings, and adobe (sun-dried brick) buildings least well – as proved by the 2003 Iranian earthquake in Bam, a predominantly adobe city, which cost more than 26,000 lives. The adobe buildings of the Middle East and South America, durable and cool inside as they are, climatically speaking, are unable to withstand even 10 per cent of the acceleration due to gravity in a horizontal direction. The situation is made worse by the fact that since adobe is not strong, builders compensate by making house walls thick, which makes them heavy and, in an earthquake, lethal to their occupants.

In *Disaster Deferred*, Stein discusses how the percentage of collapsed buildings varies with earthquake intensity on the Mercalli scale (I–XII) for different construction materials. He then relates this to the New Madrid earthquake in December 1811 through the following analysis. As in any earthquake, writes Stein:

> The intensity of shaking got smaller with increasing distance from the [epicentre]. New Madrid itself experienced shaking with intensity about IX. If there had been unreinforced brick buildings there, about half would have collapsed. About 20 per cent of wood-frame houses would have collapsed as well as 10 per cent of reinforced concrete buildings (which hadn't been invented yet). Farther from the earthquake, Memphis (which didn't exist yet) would have experienced shaking with intensity about VII, which would have collapsed about 5 per cent of unreinforced brick buildings, but few if any wood-frame or concrete buildings. Even farther away, the shaking at St Louis (which did exist) was intensity VI, and buildings didn't collapse.[31]

The point of such figures, for Stein, is to quantify honestly the risk to the American Midwest from another severe earthquake in the region. In his view, as the data from GPS measurements of local plate movement become more extensive over time, the less likely a severe earthquake in the area appears to be; and the fewer grounds there are for the 'better safe than sorry' argument of the US government. Its Federal Emergency Management Agency (FEMA), supported by the USGS, has been pushing for building codes in the Midwest as stringent as those in California – despite their predictably astronomical financial consequences, which would have to be borne by the cities in question, requiring cuts in other urban services. (Even in California, most of the state's hospitals do not meet the standards for seismic retrofits, which would cost more than $50 billion.) Stein is unconvinced both by the risk of such a Midwest earthquake recurring and by the proposed federal antidote. 'The reason there's been little discussion of the costs of stringent construction standards is the assumption that someone else will pay', he writes.[32] He calls the FEMA proposal 'an expensive cure for the wrong disease', like 'chemotherapy for a cold'.[33]

How likely is a repeat, high-magnitude, intraplate earthquake in Missouri during the next few decades? As we know, plate-tectonic theory does not predict any major earthquakes in the middle of plates; and this is supported by the dearth of plate movement in the Midwest as measured by GPS surveys. No major earthquake has occurred there for two centuries, unlike in the area of the San Andreas fault; instead there have been midwestern aftershocks of low magnitude, especially during the past century. Some palaeoseismic observations of sand blows suggest that there were previous large earthquakes in the Midwest in 900 and 1450; the record is admittedly rather meagre compared with the San Andreas excavations at Pallett Creek in California, for lack of a surface fault in the New Madrid zone to excavate. So the answer to the '$64,000 question' seems to be that such an event is highly unlikely. California is justified in spending heavily on protecting itself against severe earthquakes; Missouri and the Midwest are not. The buildings of New Madrid are more likely to fall down from natural decay than to collapse during an earthquake. This, at least, is Stein's informed but dissenting opinion, after three or four decades of work in the field.

Back in earthquake country, such as California, many home-owners who can afford to do so will continue to retrofit their properties, by bolting their houses to their foundations, rebuilding chimneys and installing automatic shutoff valves, which in the event of a rupture of a gas line should prevent houses from going up in flames – as happened in the Great Kanto earthquake in Tokyo. Still simpler precautions, which are virtually cost-free, can make the difference between escape and injury, or worse. Heavy items, such as furniture and refrigerators, should be attached or strapped to wall studs. There is a one-in-three chance that an earthquake will strike when people are asleep. So it makes sense to avoid having objects adjacent to the bed that could fall on it in a mere second if the house begins to shake. In other words, speaking for myself, were I living in shake-prone San Francisco instead of in a flat in generally very stable London, I would be foolish to continue with my habit of storing piles of heavy earth science books on the shelf directly above the bed.

Given such seismic awareness around the house, one might expect earthquakes to be a staple of conversation in California. But in reality, most Californians, like most Japanese, do not dwell on the subject. For whatever reason, the overwhelming 'policy of assumed indifference' of San Franciscans in the aftermath of the 1906 earthquake and fire (discussed in Chapter 6) predominates in today's California, too, except among seismologists, geophysicists, engineers, architects, urban planners and insurance underwriters.

California has never lacked for cults, yet there is no Californian earthquake cult and curiously little Californian earthquake culture. Mark Twain experienced a severe shock there in 1865, was respectfully impressed, and reported his reaction for the local press, including a spoof of pointless earthquake forecasting along the lines of weather forecasting, such as: 'Oct. 25. – Occasional shakes, followed by light showers of bricks and plastering. N.B. – Stand from under.'[34] F. Scott Fitzgerald's novel, *The Last Tycoon*, includes an earthquake in a Hollywood studio that brings the major characters together. In pop music, The Grateful Dead referred to the 1989 San Francisco quake in one of their less memorable recordings, 'California Earthquake (A Whole Lotta Shakin' Goin' On)'; but The Beach Boys' most celebrated song, 'Good Vibrations', contains no hint of an earthquake.

More surprisingly, in a state that draws justified attention to its scenery, the government makes scarcely any attempt to signpost the San Andreas fault for tourists as a natural feature as important in its own way as, say, the Grand Canyon in Arizona; only three such displays exist over the fault's entire length. When journalist Philip Fradkin – a longtime resident of California – travelled the San Andreas in the 1990s and wrote about his experiences in his lively account, *Magnitude 8*, he asked people: 'What does it feel like to live near, or on top of such a powerful force?' His frank summary of their responses was:

> Most people are unaware of its existence, since there is an almost total absence of signage, and damaging events are infrequent. Others give very little thought to its presence. These were the prevalent attitudes that I encountered along the fault line, where the transient nature of the populace does not lend itself to long-term memories.[35]

The film producers of Hollywood, alert as ever to making money from the public mood, must have sensed this public indifference, too. There are three disaster movies about earthquakes in California – *Earthquake* (1974), *Escape from L.A.* (1996) and *San Andreas* (2015) – but that is about all on earthquakes from Hollywood during the past century of feature films. Closer to reality, in the theme park of Hollywood's Universal Studios (which produced *Earthquake*), the long-established attraction known as 'Earthquake: The Big One' is firmly situated in San Francisco, not Los Angeles. And on Sunset Boulevard, which lies either directly on top of the Hollywood fault or just north of it, there is no sign whatsoever to indicate the fault's ghostly presence. 'It's not an attraction', according to the tourist information centre run by the Los Angeles Convention and Visitors Bureau. With so much at stake, and so little certainty about when the 'Big One' will strike Los Angeles, it is only human to deny – or at least try to forget – the continual risks of cohabiting with the San Andreas fault system. Despite the grim warning offered by the Northridge earthquake in 1994 (in which the San Andreas fault slipped a mere 1 metre, or 3 feet), little has been done to retrofit the high-rise buildings of Los Angeles against a major earthquake because of the high costs involved and a consequent lack of political will – unlike the retrofitting of buildings in Tokyo following the Kobe earthquake in 1995.

This neglect of future earthquakes by the general public is not simple to explain, especially when we recall that archaeologists and historians, too, generally neglect earthquakes. Great earthquakes tend to be forgotten, subsumed into the vicissitudes of human behaviour and experience, such as wars, economic cycles, epidemics and environmental abuses. Perhaps only the San Francisco earthquake in 1906, of all the great earthquakes in history, is widely remembered around the world.

One reason must be that large earthquakes are infrequent events. The overwhelming majority of the world's population will – thankfully – never experience such an earthquake. Even in California, the probability of an earthquake of magnitude 7.5 or larger striking somewhere in the state in any given year is estimated to be about 2 per cent: too low to become an everyday concern for residents.

Less obviously, those known to be at substantial risk, for instance people living in and around San Francisco and Los Angeles, Tokyo and many parts of Japan and on the coast of Chile, presumably suppress their awareness of earthquakes out of anxiety at their helplessness. For great earthquakes, unlike other forces of nature – floods, fires, volcanoes, hurricanes and even tornadoes and tsunamis – strike without more than minimal warning and often without any effective protection. Even in the 21st century, they are still essentially 'acts of God', in other words wholly beyond human control.

And this is perhaps a clue to the main reason that humans prefer not to think too much about them. We instinctively view ourselves as free agents, in control of our own destinies, not as bound victims of the forces of nature. As anthropologist Edward Simpson muses about the 2001 Gujarat earthquake and its survivors:

> We are unable to comprehend the enormity of the event. When we realise, and we do realise, that we have reached the limits of our capacity to comprehend, we make smaller earthquakes in our mind. These lesser earthquakes make us happy because we have created them and can comprehend them. In the process, the true earthquake, the one that shook things on the ground and killed people, is lost from view.[36]

Surely Murakami would agree. None of his stories in *after the quake* takes place in Kobe itself. Their drama resides not in the devastation of the real earthquake, but rather in the disturbing mental tremors that the event creates in people elsewhere in Japan, such as a Tokyo child who has nightmares about being stuffed into a small box by a mysterious old Earthquake Man.

Understandable as it is, such amnesia makes for poor history. As this book has shown, these earth-shattering events have played an influential and fascinating part in the politics, economics and culture of many nations and regions of the world from antiquity to the present day. By studying their history, modern civilization – rather than continuing to exist 'by geological consent, subject to change without notice' – can learn how to coexist more securely and creatively with seismic hazard.[37]

APPENDIX: CHRONOLOGY OF EARTHQUAKES

This list includes only the most lethal, most destructive or otherwise significant earthquakes mentioned in the book, plus some other notably severe earthquakes. Magnitudes are omitted, since accurate magnitude figures did not become available until the mid-20th century.

Year	**Region and/or city affected**
BC	
1831	Shandong, China
c. 1200	Eastern Mediterranean
c. 464	Sparta, Greece
461	Rome, Italy
31	Jericho, Palestine
AD	
62/63	Bay of Naples, Italy, including Pompeii
115	Antioch, Turkey
138	Rosei, China
363	Eastern Mediterranean, including Sicily, Constantinople and Jerusalem
458	Antioch, Turkey
526	Antioch, Turkey
551	Lebanon
856	Corinth, Greece
869	Sanriku coast, Japan
1138	Aleppo, Syria
1290	Chihli, China
1349	Central Italy, including Rome
1531	Lisbon, Portugal
1556	Shaanxi, China
1611	Sanriku coast, Japan
1692	Port Royal, Jamaica
1693	Catania, Italy
1746	Lima, Peru
1750	England, including London
1755	Lisbon, Portugal
1755	Cape Ann and Boston, Massachusetts, USA
1783	Calabria, Italy
1793	Sanriku coast, Japan
1811	New Madrid, Missouri, USA
1812	Venezuela, including Caracas
1819	Rann of Kutch, India
1835	Concepción, Chile
1855	Edo (Tokyo), Japan
1856	Sanriku coast, Japan
1857	Fort Tejon, California, USA
1857	Basilicata, Italy
1868	Hayward, California, USA
1880	Tokyo and Yokohama, Japan
1884	Colchester, United Kingdom
1886	Charleston, South Carolina, USA
1891	Mino-Owari, Japan
1896	Sanriku coast, Japan
1897	Assam, India
1906	San Francisco, California, USA
1908	Messina, Italy
1912	Mürefte, Turkey
1915	Avezzano, Italy
1920	Haiyuan/Kansu, China
1923	Kanto, Japan, including Tokyo and Yokohama
1927	Jericho, Palestine
1933	Sanriku coast, Japan
1933	Long Beach, California, USA
1934	Nepal and Bihar, India
1935	Quetta, Pakistan
1939	Erzincan, Turkey
1944	San Juan, Argentina
1949	Gharm Oblast, Tajikistan
1950	Assam, India and Tibet
1956	Anjar, India
1960	Agadir, Morocco
1960	Chile
1964	Prince William Sound, Alaska, USA
1970	Ancash, Peru
1971	San Fernando, California, USA
1972	Managua, Nicaragua
1975	Haicheng, China
1976	Guatemala
1976	Tangshan, China
1977	Vrancea, Romania
1980	El Asnam, Algeria
1980	Southern Italy
1985	Michoacan, Mexico, including Mexico City
1988	Northern Territory, Australia
1988	Spitak, Armenia
1989	Loma Prieta, California, USA
1990	Caspian Sea, Iran
1990	Luzon, Philippines
1992	Landers, California, USA
1992	Cairo, Egypt
1993	Latur, India
1994	Northridge, California, USA
1995	Kobe, Japan
1998	Papua New Guinea
1999	Izmit, Turkey
2001	Gujarat, India
2003	Bam, Iran
2004	Sumatra, Indonesia and Indian Ocean
2005	Kashmir, Pakistan
2008	Sichuan, China
2009	L'Aquila, Italy
2010	Port-au-Prince, Haiti
2010	Chile
2010	Canterbury, New Zealand
2011	Tohoku, Japan
2012	Sumatra, Indonesia and Indian Ocean
2015	Nepal

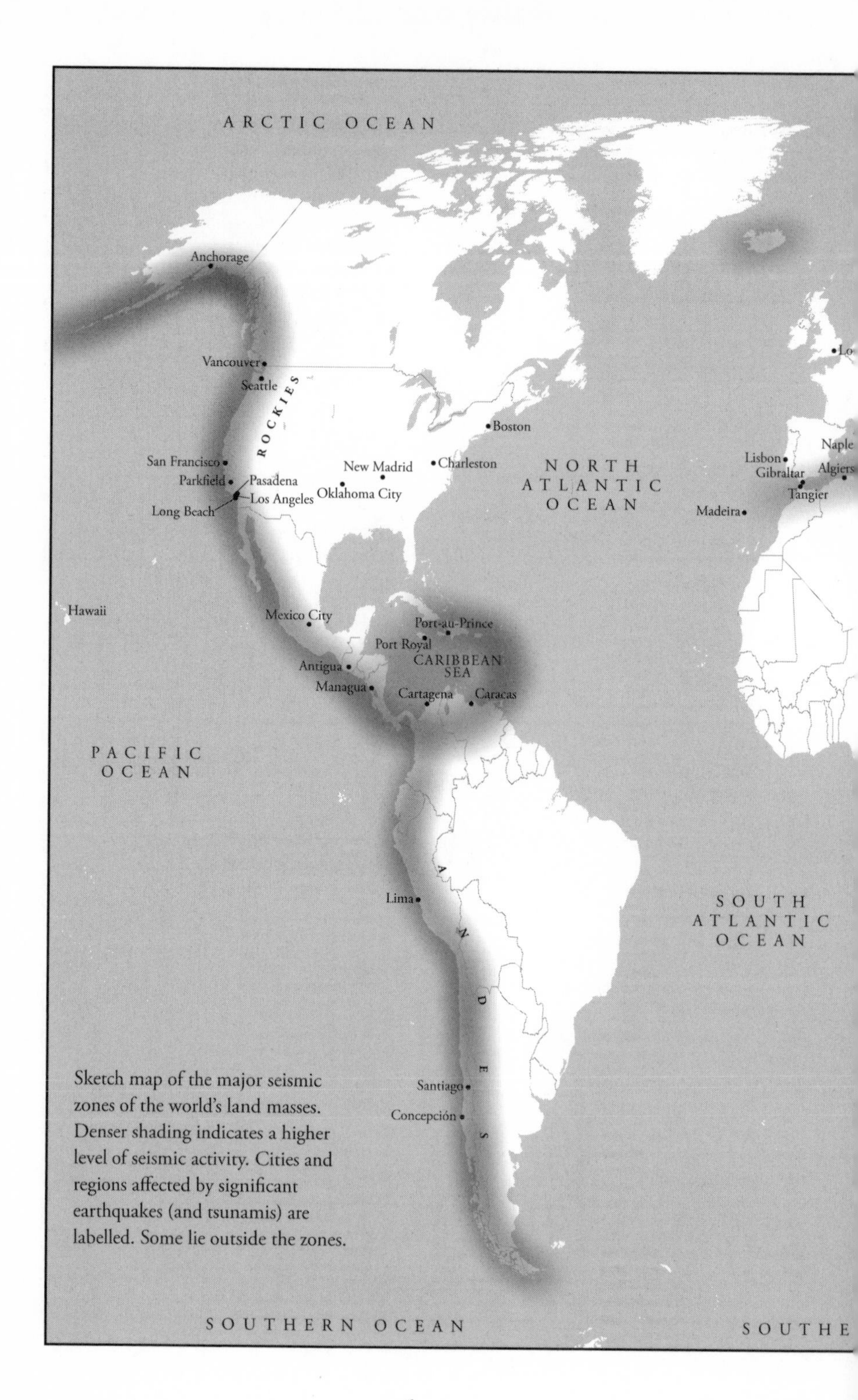

Sketch map of the major seismic zones of the world's land masses. Denser shading indicates a higher level of seismic activity. Cities and regions affected by significant earthquakes (and tsunamis) are labelled. Some lie outside the zones.

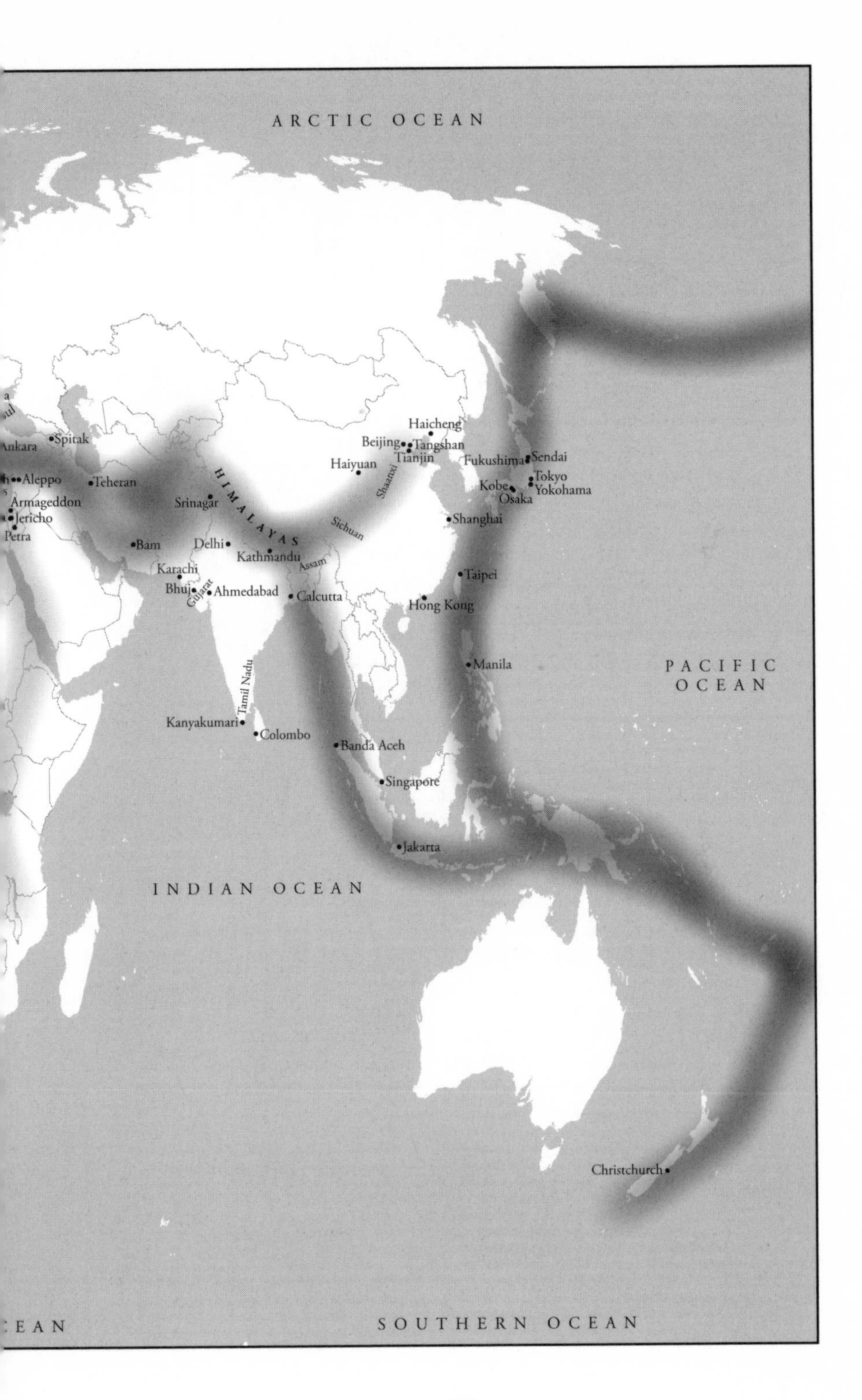
ARCTIC OCEAN
Haicheng
Beijing
Tangshan
Tianjin
Haiyuan
Fukushima
Sendai
Tokyo
Yokohama
Kobe
Osaka
Spitak
Ankara
Aleppo
Teheran
Armageddon
Jericho
Petra
Srinagar
HIMALAYAS
Shaanxi
Sichuan
Shanghai
Bam
Delhi
Kathmandu
Assam
Karachi
Bhuj
Gujarat
Ahmedabad
Calcutta
Taipei
Hong Kong
Manila
PACIFIC OCEAN
Tamil Nadu
Kanyakumari
Colombo
Banda Aceh
Singapore
Jakarta
INDIAN OCEAN
Christchurch
SOUTHERN OCEAN

NOTES AND REFERENCES

Full publication details for works cited in abbreviated form below can be found in the Bibliography.

INTRODUCTION: EARTHQUAKES AND HISTORY

1 McNutt: 1397.
2 McGarr et al.: 830.
3 Quoted in Galchen: 38–39.
4 Quoted in Mike Soraghan, 'Oklahoma Agency Linked Quakes to Oil in 2010, but Kept Mum Amid Industry Pressure', 3 Mar. 2015, http://www.eenews.net/stories/1060014342.
5 Witze, 'Artificial Quakes Shake Oklahoma': 419.
6 Quoted in Mike Soraghan, 'Oklahoma Agency Linked Quakes to Oil in 2010, but Kept Mum Amid Industry Pressure', 3 Mar. 2015, http://www.eenews.net/stories/1060014342.
7 Hough and Bilham: 7.
8 Darwin: 232.
9 William Shakespeare, *Romeo and Juliet*: act 1, scene iii.
10 Haining: 86.
11 Smith: 24.
12 'Earthquakes, Tsunamis, and the Related Vulnerability in South America and the Caribbean – An Overview', in Ismail-Zadeh, Fucugauchi, Kijko, Takeuchi and Zaliapin eds.: 146.
13 Quoted in Herbert-Gustar and Nott: 133.
14 Rozario: 3.
15 Quoted in Kendrick: 132.
16 M. K. Gandhi, 'Superstition v. Faith', *The Collected Works of Mahatma Gandhi* (New Delhi, 1958–), vol. 63: 165.
17 Jackson: 1911.
18 Attributed to Durant in *The Oxford Dictionary of American Quotations*, Hugh Rawson and Margaret Miner eds. (New York, 2005): 600.
19 Nur with Burgess: 228.
20 Ibid.: 107.
21 Ibid.: 259.
22 Quoted in Tania Branigan, 'Earthquake and Tsunami "Japan's Worst Crisis Since Second World War"', *Guardian*, 14 March 2011.

1 EARTHQUAKES BEFORE SEISMOLOGY

1 Quoted in W. H. D. Rouse, *The Story of Achilles: A Translation of Homer's 'Iliad' into Plain English* (London, 1938): 389.
2 Matthew, 27: 50–52; 28: 2, *The New English Bible* (Oxford and Cambridge, 1970).
3 1 Kings, 19: 11–12, quoted in Nur with Burgess: 85.
4 Nur with Burgess: 85.
5 Genesis, 13: 13, *The New English Bible* (Oxford and Cambridge, 1970).
6 Ibid., 19: 24–25, 28.
7 Strabo: 297.
8 Nur with Burgess: 219.
9 Ibid.: 235.
10 Drews: 38.
11 Diodorus Siculus: 289.
12 Cartledge: 26.
13 Quoted in Heiken, Funiciello and De Rita: 99.
14 Heiken, Funiciello and De Rita: 97.
15 Quoted in Heiken, Funiciello and De Rita: 100.
16 Mallet, vol. 2: 157–58.
17 Quoted in Roberts: 274.
18 Pliny the Younger, *The Letters of Pliny the Younger*, Betty Radice trans. (London, 1963): 170.
19 Roberts: 274.
20 Edward Gibbon, *The History of the Decline and Fall of the Roman Empire*, David Womersley ed. (London, 2000): 279.
21 Quoted in Palmer: 2.
22 Smits, 'Shaking Up Japan': 1046.
23 Ibid.: 1072.

2 THE YEAR OF EARTHQUAKES: LONDON, 1750

1 Stukeley: 732. The dates of the earthquakes in this chapter refer to the Julian ('Old Style') calendar, rather than the Gregorian ('New Style') calendar adopted in England in 1752 and still in use today. In the Gregorian calendar, the two London earthquakes occurred on 19 February and 19 March 1750.
2 Walpole: 193.
3 Ibid.: 198–99.
4 Folkes: 615.
5 Davison: 335.
6 Walpole: 199.
7 Ibid.: 201.
8 Ibid.: 202–3.
9 *Gentleman's Magazine*, 20 (1750): 185.
10 Davison: 336.
11 Walpole: 200.
12 Quoted in Dvorak: 18.
13 John and Charles Wesley, *Hymns Occasioned by the Earthquake, March 8, 1750*, 2nd edn (Bristol: 1756).
14 Quoted in Kendrick: 1.
15 Quoted in Kendrick: 9.
16 Quoted in Kendrick: 12.
17 Quoted in *Gentleman's Magazine*, 20 (1750): 123.
18 Quoted in Kendrick: 5.
19 Quoted in Kendrick: 20.
20 Kendrick: 21.
21 Davison: 2.
22 *Gentleman's Magazine*, 20 (1750): 89.
23 Quoted in Willmoth: 25.
24 Willmoth: 28.
25 Walpole: 207.
26 Stukeley: 739.
27 Ibid.: 738.
28 Ibid.: 745.
29 Quoted in Bolt: 8.
30 Stukeley: 745.
31 Walpole: 207.
32 Quoted in Robert G. Ingram, 'Earthquakes, Religion and Public Life in Britain during the 1750s', in Braun and Radner eds.: 115.

3 THE WRATH OF GOD: LISBON, 1755

1 Coen: 107.
2 *Illustrated London News* (30 Mar. 1850): 222.
3 Charles Dickens, 'Lisbon', *Household Words* (25 Dec. 1858): 89.
4 Paice: xvi.
5 Gould: 402.
6 Kendrick: 29.
7 Quoted in Maxwell: 17.
8 C. R. Boxer, *The Portuguese Seaborne Empire, 1415–1825* (London, 1977): 189.
9 Quoted in Paice: 65.
10 Quoted in Paice: 73.
11 Quoted in Paice: 115–16.
12 Quoted in Paice: 131.
13 Hough and Bilham: 42.
14 Quoted in Paice: 82.
15 Charles Davison, *Great Earthquakes* (London, 1936): 3.
16 Quoted in Maxwell: 2.
17 Maxwell: 2.
18 Quoted in Kendrick: 45.
19 Maxwell: 24.
20 Quoted in Paice: 192.
21 Alexander Pope, *An Essay on Man*, Epistle 1, ll. 285–94.
22 Voltaire, *Candide and Other Stories*, Roger Pearson trans. (Oxford, 2006): 13.
23 Quoted in Kendrick: 149.
24 Quoted in Maxwell: 20.
25 Rozario: 17–18.
26 'Lisbon', *Household Words* (25 Dec. 1858): 88.

4 BIRTH OF NATIONS: CARACAS, 1812

1 Quoted in Lynch: 1.
2 Quoted in Harvey: 82–83.
3 Rogelio Altez, 'New Interpretations of the Social and Material Impacts of the 1812 Earthquake in Caracas, Venezuela', in Sintubin, Stewart, Niemi and Altunel eds.: 49.
4 Humboldt and Bonpland: 451.
5 Quoted in Arana: 109.

6 Arana: 110.
7 Quoted in Arana: 122.
8 Quoted in Arana: 126.
9 Harvey: 91.
10 Ibid.: 97.
11 Lynch: 64.
12 Bolívar: 3.
13 Ibid.: 7–8.
14 Quoted in Arana: 59.
15 Bolívar: 6.
16 Ibid.: 5.
17 Quoted in Hough, *Earthshaking Science*: 67.
18 Hough, *Earthshaking Science*: 67.
19 Humboldt and Bonpland: 449.
20 Quoted in Hough and Bilham: 54.
21 Hough and Bilham: 55.
22 Quoted in Zeilinga de Boer and Sanders: 129.
23 Humboldt and Bonpland: 473.
24 Quoted in Arana: 74.
25 Quoted in Arana: 175.
26 Bolívar: 7.
27 Arana: 5.
28 Introduction to Bolívar: xliii.

5 SEISMOLOGY BEGINS: NAPLES, 1857

1 Quoted in Walker: 50.
2 Mallet, vol. 1: vii–viii.
3 Ibid.: ix.
4 Ferrari and McConnell: 51.
5 Quoted in Ferrari and McConnell: 51–52.
6 Mallet, vol. 2: 3.
7 Ibid., vol. 1: 35–36.
8 Ibid.: 410–11.
9 Hough and Bilham: 93.
10 Ferrari and McConnell: 62.
11 Mallet, vol. 2: 380–81.
12 Dewey and Byerly: 195.
13 Quoted in Herbert-Gustar and Nott: 71.
14 Quoted in Herbert-Gustar and Nott: 52.
15 Quoted in Herbert-Gustar and Nott: 58.
16 Quoted in Clancey: 64.
17 Quoted in Clancey: 65.
18 Clancey: 101.
19 Quoted in Herbert-Gustar and Nott: 91.
20 Quoted in Talwani: 1372.
21 Quoted in Talwani: 1370.
22 Quoted in Talwani: 1371.
23 Talwani: 1372.
24 *San Francisco Call*, 5 Aug. 1906.

6 ELASTIC REBOUND: SAN FRANCISCO, 1906

1 Quoted in Fradkin: 102.
2 Hough and Bilham: 244.
3 Quoted in Hansen and Condon: 15.
4 Quoted in Hansen and Condon: 15–16.
5 Quoted in Hansen and Condon: 27.
6 Quoted in Hansen and Condon: 27.
7 Quoted in Hansen and Condon: 33, 36.
8 James: 1215–16.
9 Ibid.: 1218.
10 Quoted in Hansen and Condon: 70.
11 Hansen and Condon: 94.
12 Ibid.: 27.
13 Quoted in Fradkin: 136.
14 Quoted in Fradkin: 81.
15 Quoted in Hansen and Condon: 111.
16 Quoted in Hansen and Condon: 110.
17 Quoted in Hansen and Condon: 108–9.
18 Hansen and Condon: 108–9.
19 Quoted in Hansen and Condon: 124.
20 Quoted in Hansen and Condon: 124.
21 Hansen and Condon: 127.
22 *New York Times*, 23 Apr. 1906.
23 Coen: 224.
24 Odell and Weidenmier: 1024.
25 Quoted in Hansen and Condon: 135.
26 Hansen and Condon: 135.
27 Thomas C. Hanks and Helmut Krawinkler, 'The 1989 Loma Prieta Earthquake and Its Effects: Introduction to the Special Issue', *Bulletin of the Seismological Society of America*, 81 (1991): 1420–21.
28 Richter: 498.
29 McWilliams: 41–42. The article was first published in *American Mercury*, 29 (1933): 199–201.
30 Quoted in Fradkin: 120.

7 HOLOCAUST IN JAPAN: TOKYO AND YOKOHAMA, 1923

1 Quoted in Zeilinga de Boer and Sanders: 185.
2 [Bureau of Social Affairs], *Great Earthquake*: 137.
3 Kurosawa: 50.
4 Ibid.: 52–54.
5 Clancey: 218.
6 Quoted in Clancey: 218.
7 Quoted in Clancey: 220.
8 Quoted in Clancey: 220.
9 *The Age* (Melbourne), 4 Sept. 1923.
10 Quoted in Bolt: 20.
11 Clancey: 221.
12 Quoted in Hadfield: 2–3.
13 Quoted in Hadfield: 3.
14 Hadfield: 5 (based on reports in the *Japan Times* during Sept. 1923).
15 Waley: 171–72.
16 Akutagawa: 197.
17 Quoted in Seidensticker: 39.
18 Kawabata: 105–8.
19 Quoted in Schencking: 307.
20 Quoted in Schencking: 307.
21 Quoted in Schencking: 282.
22 [Bureau of Social Affairs], *Great Earthquake*: 33.
23 Weisenfeld: 310 (footnote 7).
24 Seidensticker: 99.
25 Samuels: 55.
26 Pilling: 72.
27 Ibid.: 304.
28 Hammer: 259.
29 Seidensticker: 121.

8 BIRTH PANG OF A NEW CHINA: TANGSHAN, 1976

1 Hough, *Predicting the Unpredictable*: 80.
2 Ibid.
3 Qian Gang: 279.
4 Ibid.: 301.
5 Quoted in Qian Gang: 309.
6 Chen: 236.
7 Quoted in Chen: 235.
8 Qian Gang: 53.
9 Quoted in Qian Gang: 59.
10 Quoted in Qian Gang: 60–61.
11 Palmer: 171.
12 Quoted in Qian Gang: 143.
13 Quoted in Palmer: 172.
14 Palmer: 145, 147.
15 Qian Gang: 227.
16 Quoted in Palmer: 109.
17 Quoted in Palmer: 166.
18 Chen: 240.
19 Qian Gang: 200.
20 Ibid.: 206.
21 Quoted in Qian Gang: 206.
22 Qian Gang: 207.
23 Palmer: 163.
24 Quoted in Qian Gang: 129.
25 Quoted in Palmer: 132.
26 Quoted in Chen: 242–43.
27 Quoted in Palmer: 172.
28 Palmer: 197.
29 Ibid.
30 Quoted in Qian Gang: 267–68.
31 Quoted in Palmer: 191.
32 Quoted in Palmer: 189.
33 Palmer: 209.
34 Chen: 244
35 Palmer: 247
36 Quoted in Qian Gang: 232.
37 Frank Dikötter, 'Number Two Capitalist Roader', *Literary Review*, 452 (June 2015): 6. (This is a review of Alexander V. Pantsov and Steve I. Levine, *Deng Xiaoping: A Revolutionary Life*, a book that briefly discusses Deng's response to the Tangshan earthquake.)

9 GRIEF AND GROWTH IN THE LAND OF GANDHI: GUJARAT, 2001

1 Prantik Mandal and M. Rodkin, 'Spatiotemporal Variation of Fractal Properties in the Source Zone of the 2001 *Mw* 7.7 Bhuj Earthquake', *Bulletin of the Seismological Society of America*, 104 (2014): 2060.
2 Robert L. Kovach, Kelly Grijalva and Amos Nur, 'Earthquakes and Civilizations of the Indus: A Challenge for Archaeoseismology', in Sintubin,

Stewart, Niemi and Altunel eds.: 121.

3 James Burnes, *Narrative of a Visit to the Court of Sinde at Hyderabad on the Indus: With a Sketch of the History of Cutch* (Edinburgh and London, 1839): 66–67.
4 Simpson: 237.
5 Ibid.: 236.
6 Quoted in Simpson: 241.
7 Quoted in Simpson: 241.
8 Simpson: 242.
9 Ibid.: 59.
10 Ibid.: 6.
11 Victor Mallet, 'India: Narendra Modi's Market Model', *Financial Times*, 5 Feb. 2014.
12 Quoted in Victor Mallet, 'India: Narendra Modi's Market Model', *Financial Times*, 5 Feb. 2014.
13 Maitreesh Ghatak and Sanchari Roy, 'Did Gujarat's Growth Rate Accelerate under Modi?', *Economic & Political Weekly*, 12 Apr. 2014: 12.
14 Ibid.: 15.
15 Quoted in Jinoy Jose P, 'Disruption Is Needed After Disasters', *The Hindu*, 22 Sept. 2014.
16 Simpson: 13.
17 Quoted in Preeti Panwar, 'Why Did Narendra Modi Choose Bhuj to Give His Independence Day Speech?', 16 Aug. 2013, http://www.oneindia.com/feature/2013/why-did-narendra-modi-choose-bhuj-to-give-his-independence-day-speech-1284693.html.
18 Preeti Panwar, 'Why Did Narendra Modi Choose Bhuj to Give His Independence Day Speech?', 16 August 2013, http://www.oneindia.com/feature/2013/why-did-narendra-modi-choose-bhuj-to-give-his-independence-day-speech-1284693.html.
19 Quoted in 'Modi Shares Nepal's Grief', *The Hindu*, 27 Apr. 2015.
20 Patel and Revi, eds.: 401.
21 Bilham: 46.
22 Patel and Revi, eds.: 391.
23 Ibid.: 393.
24 Gautam Bhatia, 'The Great Delhi Earthquake of 2017', *Outlook*, 12 May 2015.

10 WAR AND PEACE BY TSUNAMI: THE INDIAN OCEAN, 2004

1 Stein: 100.
2 Titov, Rabinovich, Mofjeld, Thomson and González: 2047.
3 Nield: 262–63.
4 Deraniyagala: 32–33.
5 Quoted in Karan and Subbiah, eds.: 241.
6 Quoted in Pisani: 176–77.
7 Interview with Mahdi in *Stories of Recovery, 10 Years after the Tsunami* (2014), a film by Jolyon Hoff, at: https://www.youtube.com/watch?v=fXpBgdAon9E.
8 Pisani: 249.
9 Christopher Jasparro and Jonathan Taylor, 'Transnational Geopolitical Competition and Natural Disasters: Lessons from the Indian Ocean Tsunami', in Karan and Subbiah, eds.: 283.
10 Quoted in Hyndman: 26.
11 Quoted in Hyndman: 26.
12 Quoted in Hyndman: 114.
13 Pisani: 229.
14 Hyndman: 15.
15 Ibid.: 28.
16 Harrison: 227.
17 Witze, 'Tsunami Alerts Fall Short': 152.

11 MELTDOWN AND AFTER: FUKUSHIMA, 2011

1 Kenji Satake, 'The 2011 Tohoku, Japan, Earthquake and Tsunami', in Ismail-Zadeh, Fucugauchi, Kijko, Takeuchi and Zaliapin eds.: 311.
2 Ibid.: 316.
3 Lafcadio Hearn, *Gleanings in Buddha-Fields: Studies of Hand and Soul in the Far East* (London, 1897): 16. The article originally appeared in the *Atlantic Monthly* in Dec. 1896.
4 Smits, *When the Earth Roars*: 33–34.

5 Ibid.: 119.
6 Pilling: 6.
7 Quoted in Smits, *When the Earth Roars*: 123.
8 Musson: 139.
9 Personal information from Pico Iyer, whose anonymous report appears at http://dalailama.com/news/post/768-the-transformation-of-pain.
10 Quoted in Tania Branigan, 'Earthquake and Tsunami "Japan's Worst Crisis Since Second World War"', *Guardian*, 14 March 2011.
11 Lochbaum, Lyman and Stranahan: 11–12.
12 Quoted in Lochbaum, Lyman and Stranahan: 94.
13 Lochbaum, Lyman and Stranahan: 109.
14 Stein and Stein: 27.
15 Quoted in Lochbaum, Lyman and Stranahan: 53.
16 Lochbaum, Lyman and Stranahan: 51.
17 Samuels: 38.
18 Smits, *When the Earth Roars*: 23.
19 Quoted in Samuels: 131.
20 Samuels: 191.
21 Quoted in Kieffer: 262.
22 Samuels: 52.
23 Quoted in Samuels: x.
24 Quoted in Samuels: 180.

CONCLUSION: EARTHQUAKES, NATIONS AND CIVILIZATION

1 *Aftershock* was released by Metrodome in December 2010 in Mandarin with English subtitles.
2 Qian Gang: 22.
3 Vale and Campanella eds.: 347.
4 Anselm Smolka, 'Extreme Geohazards: Risk Management from a (Re)insurance Perspective', in Ismail-Zadeh, Fucugauchi, Kijko, Takeuchi and Zaliapin eds.: 367.
5 Hough, *Earthshaking Science*: 123.
6 Quoted in Herbert-Gustar and Nott: 163–64.
7 Richter: 386–87.
8 Quoted in Hough, *Richter's Scale*: 265.
9 Musson: 165.
10 Introduction to Frank Press, 'Earthquake Prediction', *Scientific American*, 232 (1975): 14.
11 Hough, *Predicting the Unpredictable*: 110.
12 Quoted in Tributsch: 15.
13 Hough, *Predicting the Unpredictable*: 126.
14 Musson: 153.
15 Seth Stein, 'Seismic Gaps and Grizzly Bears', *Nature*, 356 (1992): 388.
16 William Robbins, 'Midwest Quake Is Predicted; Talk Is Real', *New York Times*, 20 Aug. 1990.
17 Quoted in Richard A. Kerr, 'The Lessons of Dr Browning', *Science*, 253 (1991): 622.
18 Quoted in Robinson, *Earthshock*: 74.
19 Jacob M. Appel, 'A Comparative Seismology', *Weber*, 18 (2001): 92.
20 Stein: 16.
21 Quoted in Olson: 137.
22 Quoted in Robinson, *Earthshock*: 75.
23 Ambraseys and Bilham: 153.
24 Musson: 238.
25 Quoted in Hadfield: 187–88.
26 Hadfield: 187.
27 Murakami: 2.
28 Ibid.: 17.
29 Gina L. Barnes, 'Earthquake Archaeology in Japan: An Overview', in Sintubin, Stewart, Niemi and Altunel eds.: 86.
30 Hough, *Predicting the Unpredictable*: 217.
31 Stein: 225–26.
32 Ibid.: 228.
33 Ibid.: 234.
34 Quoted in Coen: 58.
35 Fradkin: 11.
36 Simpson: 250.
37 Attributed to Will Durant in *The Oxford Dictionary of American Quotations*, Hugh Rawson and Margaret Miner eds. (New York, 2005): 600.

BIBLIOGRAPHY

Akutagawa, Ryunosuke, *Rashomon and Other Stories*, Jay Rubin trans. (London, 2006)

Ambraseys, N. N., 'Value of Historical Records of Earthquakes', *Nature*, 232 (1971): 375–79

Ambraseys, Nicholas, and Roger Bilham, 'Corruption Kills', *Nature*, 469 (2011): 153–55

Arana, Marie, *Bolívar: American Liberator* (London, 2013)

Bilham, Roger, 'Subterranean Shifts: The Science Behind Earthquakes in the Himalaya', *Himal*, 28 (2015): 44–58

Bolívar, Simón, *El Libertador: Writings of Simón Bolívar*, Frederick H. Fornoff trans., David Bushnell ed. (New York, 2003)

Bolt, Bruce A., *Earthquakes and Geological Discovery* (New York, 1993)

Braun, Theodore E. D., and John B. Radner, eds., *The Lisbon Earthquake of 1755: Representations and Reactions* (Oxford, 2005)

Brett, William Bailie, *A Report on the Bihar Earthquake* (Patna, 1935)

[Bureau of Social Affairs, Home Office, Japan], *The Great Earthquake of 1923 in Japan* (Tokyo, 1926)

Cartledge, P. A., 'Seismicity and Spartan Society', *Liverpool Classical Monthly*, 1 (1976): 25–28

Chen, Beatrice, '"Resist the Earthquake and Rescue Ourselves": The Reconstruction of Tangshan After the 1976 Earthquake', in Lawrence J. Vale and Thomas J. Campanella, eds., *The Resilient City: How Modern Cities Recover from Disaster* (New York, 2005): 235–54

Clancey, Gregory, *Earthquake Nation: The Cultural Politics of Japanese Seismicity, 1868–1930* (Berkeley CA, 2006)

Cline, Eric H., *1177 B.C.: The Year Civilization Collapsed* (Princeton NJ, 2014)

Coen, Deborah R., *The Earthquake Observers: Disaster Science from Lisbon to Richter* (Chicago IL, 2013)

Collier, Michael, *A Land in Motion: California's San Andreas Fault* (Berkeley CA, 1999)

Darwin, Charles, *The Voyage of the Beagle*, Janet Browne and Michael Neve eds. (London, 1989)

Davis, Mike, *Los Angeles and the Imagination of Disaster* (London, 1999)

Davison, Charles, *A History of British Earthquakes* (Cambridge, 1924)

Deraniyagala, Sonali, *Wave: A Memoir of Life After the Tsunami* (London, 2013)

Dewey, James, and Perry Byerly, 'The Early History of Seismometry', *Bulletin of the Seismological Society of America*, 59 (1969): 183–227

Diodorus Siculus, *Diodorus of Sicily*, vol. 4, C. H. Oldfather trans. (Cambridge MA: 1946)

Drews, Robert, *The End of the Bronze Age: Changes in Warfare and the Catastrophe ca. 1200 B.C.* (Princeton NJ, 1993)

Dvorak, John, *Earthquake Storms: The Fascinating History and Volatile Future of the San Andreas Fault* (New York, 2014)

[Earthquake Engineering Laboratory, California Institute of Technology], *The Great Tangshan Earthquake of 1976* (Pasadena CA, 2002)

Ferrari, Graziano, and Anita McConnell, 'Robert Mallet and the "Great Neapolitan Earthquake of 1857"', *Notes and Records of the Royal Society*, 59 (2005): 45–64

Folkes, Martin, 'The President's Account of the Earthquake in London, March 8', *Philosophical Transactions of the Royal Society*, 46 (1749–50): 613–15

Force, Eric R., *Impact of Tectonic Activity on Ancient Civilizations: Recurrent Shakeups, Tenacity, Resilience, and Change* (Lanham MD, 2015)

Fradkin, Philip L., *Magnitude 8: Earthquakes and Life Along the San Andreas Fault*, pbk edn (Berkeley CA, 1999)

Galchen, Rivka, 'Weather Underground', *New Yorker*, 13 Apr. 2015: 34–40

Gamburd, Michele Ruth, *The Golden Wave: Culture and Politics After Sri Lanka's Tsunami Disaster* (Bloomington IN, 2014)

Gould, Peter, 'Lisbon 1755: Enlightenment, Catastrophe, and Communication', in *Geography and Enlightenment*, David N. Livingstone and Charles W. J. Withers eds. (Chicago IL, 1999): 399–413

Hadfield, Peter, *Sixty Seconds That Will Change the World: How the Coming Tokyo Earthquake Will Wreak Worldwide Economic Devastation*, rev. edn (London, 1995)

Haining, Peter, *The Great English Earthquake* (London, 1976)

Hammer, Joshua, *Yokohama Burning: The Deadly 1923 Earthquake and the Fire That Helped Forge the Path to World War II* (New York, 2006)

Hansen, Gladys, and Emmet Condon, *Denial of Disaster* (San Francisco CA, 1989)

Harrison, Frances, *Still Counting the Dead: Survivors of Sri Lanka's Hidden War* (London, 2012)

Harvey, Robert, *Liberators: Latin America's Struggle for Independence, 1810–1830* (London, 2000)

Heiken, Grant, Renato Funiciello and Donatella De Rita, *The Seven Hills of Rome: A Geological Tour of the Eternal City* (Princeton NJ, 2005)

Herbert-Gustar, A. L., and P. A. Nott, *John Milne: Father of Modern Seismology* (Tenterden, 1980)

Hindmarsh, Richard, ed., *Nuclear Disaster at Fukushima Daiichi: Social, Political and Environmental Issues* (New York, 2013)

Hough, Susan, *Earthshaking Science: What We Know (and Don't Know) About Earthquakes* (Princeton NJ, 2002)

Richter's Scale: Measure of an Earthquake, Measure of a Man (Princeton NJ, 2007)

Predicting the Unpredictable: The Tumultuous Science of Earthquake Prediction (Princeton NJ, 2010)

Hough, Susan Elizabeth, and Roger G. Bilham, *After the Earth Quakes: Elastic Rebound on an Urban Planet* (New York, 2006)

Humboldt, Alexander von, and Aimé Bonpland, *Personal Narrative of Travels to the Equinoctial Regions of America*, Thomasina Ross trans., vol. 1 (London, 1852)

Hyndman, Jennifer, *Dual Disasters: Humanitarian Aid After the 2004 Tsunami* (Sterling VA, 2011)

Ismail-Zadeh, Alik, Jaime Urrutia Fucugauchi, Andrzej Kijko, Kuniyoshi Takeuchi and Ilya Zaliapin, eds., *Extreme Natural Hazards, Disaster Risks and Societal Implications* (Cambridge, 2014)

Jackson, James, 'Fatal Attraction: Living with Earthquakes, the Growth of Villages into Megacities, and Earthquake Vulnerability in the Modern World', *Philosophical Transactions of the Royal Society A*, 364 (2006): 1911–25

James, William, 'On Some Mental Effects of the Earthquake', in *William James: Writings, 1902–1910* (New York, 1987): 1215–22

Karan, Pradyumna P., and Shanmugam P. Subbiah, eds., *The Indian Ocean Tsunami: The Global Response to a Natural Disaster* (Lexington KY, 2011)

Kawabata, Yasunari, *The Dancing Girl of Izu and Other Stories*, J. Martin Holman trans. (Washington DC, 1997)

Kearey, Philip, and Frederick J. Vine, *Global Tectonics* (Oxford, 1990)

Kendrick, T. D., *The Lisbon Earthquake* (London, 1956)

Kerr, Richard A., 'Weak Faults: Breaking Out All Over', *Science*, 255 (1992): 1210–12

Kieffer, Susan W., *The Dynamics of Disaster* (New York, 2013)
King, Geoffrey, and Geoff Bailey, 'Tectonics and Human Evolution', *Antiquity*, 80 (2006): 265–86
Kingston, Jeff, ed., *Natural Disaster and Nuclear Crisis in Japan: Response and Recovery after Japan's 3/11* (London, 2012)
Klein, Naomi, *The Shock Doctrine: The Rise of Disaster Capitalism* (London, 2007)
Kondoleon, Christine, *Antioch: The Lost Ancient City* (Princeton NJ, 2000)
Kovach, Robert L., and Amos Nur, 'Earthquakes and Archaeology: Neocatastrophism or Science?', *Eos*, 87 (2006): 317–19
Kurosawa, Akira, *Something Like an Autobiography*, Audie E. Bock trans., pbk edn (New York, 1983)
Lochbaum, David, Edwin Lyman, Susan Q. Stranahan, and the Union of Concerned Scientists, *Fukushima: The Story of a Nuclear Disaster* (New York, 2014)
Lynch, John, *Simón Bolívar: A Life* (New Haven CT, 2006)
McGarr, A., et al., 'Coping with Earthquakes Induced by Fluid Injection', *Science*, 347 (2015): 830–31
McGuire, Bill, *Waking the Giant: How a Changing Climate Triggers Earthquakes, Tsunamis, and Volcanoes* (Oxford, 2012)
McNutt, Marcia, 'Integrity – Not Just a Federal Issue', *Science*, 347 (2015): 1397
McPhee, John, *Assembling California* (New York, 1993)
McWilliams, Carey, 'The Folklore of Earthquakes', in Carey McWilliams, *Fool's Paradise: A Carey McWilliams Reader*, Dean Stewart and Jeannine Gendar eds. (Berkeley CA, 2001): 41–42
Mallet, Robert, *Great Neapolitan Earthquake of 1857: The First Principles of Observational Seismology*, 2 vols. (London, 1862)
Maxwell, Kenneth, *Pombal: Paradox of the Enlightenment* (Cambridge, 1995)
Murakami, Haruki, *after the quake*, Jay Rubin trans. (London, 2002)
Musson, Roger, *The Million Death Quake: The Science of Predicting Earth's Deadliest Natural Disaster* (New York, 2012)
Nield, Ted, *Supercontinent: Ten Billion Years in the Life of Our Planet* (London, 2007)
Nur, Amos, with Dawn Burgess, *Apocalypse: Earthquakes, Archaeology, and the Wrath of God* (Princeton NJ, 2008)
Odell, Kerry A., and Marc D. Weidenmier, 'Real Shock, Monetary Aftershock: The 1906 San Francisco Earthquake and the Panic of 1907', *Journal of Economic History*, 64 (2004): 1002–27
Olson, Richard S., *The Politics of Earthquake Prediction* (Princeton NJ, 1981)
Ouwehand, C., *Namazu-e and Their Themes* (Leiden, 1964)
Paice, Edward, *Wrath of God: The Great Lisbon Earthquake of 1755* (London, 2008)
Palmer, James, *The Death of Mao: The Tangshan Earthquake and the Birth of the New China*, pbk edn (London, 2013)
Patel, Shirish B., and Aromar Revi, eds., *Recovering from Earthquakes: Response, Reconstruction and Impact Mitigation in India* (New Delhi, 2010)
Pilling, David, *Bending Diversity: Japan and the Art of Survival* (London, 2014)
Pisani, Elizabeth, *Indonesia Etc.: Exploring the Improbable Nation* (London, 2014)
Poniatowska, Elena, *Nothing, Nobody: The Voices of the Mexico City Earthquake*, Aurora Camacho de Schmidt and Arthur Schmidt trans. (Philadelphia PA, 1995)
Qian Gang, *The Great China Earthquake*, Nicola Ellis and Cathy Silber trans. (Beijing, 1989)
Reisner, Marc, *A Dangerous Place: California's Unsettling Fate* (London, 2003)
Richter, Charles, *Elementary Seismology* (San Francisco CA, 1958)
Roberts, Paul, *Life and Death in Pompeii and Herculaneum* (London, 2013)
Robinson, Andrew, *Earthshock: Hurricanes, Volcanoes, Earthquakes, Tornadoes and*

Other Forces of Nature, rev. edn (London, 2002)
Earthquake: Nature and Culture (London, 2012)
Rozario, Kevin, *The Culture of Calamity: Disaster and the Making of Modern America* (Chicago IL, 2007)
Rubinstein, Justin L., and Alireza Babaie Mahani, 'Myths and Facts on Wastewater Injection, Hydraulic Fracturing, Enhanced Oil Recovery, and Induced Seismicity', *Seismological Research Letters*, 86 (2015): 1060–67
Samuels, Richard J., *3.11: Disaster and Change in Japan* (Ithaca NY, 2013)
Schencking, J. Charles, *The Great Kanto Earthquake and the Chimera of National Reconstruction in Japan* (New York, 2013)
Schulz, Kathryn, 'The Really Big One', *New Yorker*, 20 July 2015: 52–59
Seidensticker, Edward, *Tokyo Rising: The City Since the Great Earthquake* (New York, 1990)
Simpson, Edward, *The Political Biography of an Earthquake: Aftermath and Amnesia in Gujarat, India* (London, 2013)
Sintubin, Manuel, Iain S. Stewart, Tina M. Niemi and Erhan Altunel, eds., *Ancient Earthquakes* [Geological Society of America Special Paper 471] (Boulder CO, 2010)
Smith, Roff, 'The Biggest One', *Nature*, 465 (2010): 24
Smits, Gregory, 'Shaking up Japan: Edo Society and the 1855 Catfish Picture Prints', *Journal of Social History*, 39 (2006): 1045–77
When the Earth Roars: Lessons from the History of Earthquakes in Japan (Lanham MD, 2014)
Starrs, Roy, ed., *When the Tsunami Came to Shore: Culture and Disaster in Japan* (Leiden, 2014)
Stein, Seth, *Disaster Deferred: How New Science Is Changing Our View of Earthquake Hazards in the Midwest* (New York, 2010)
Stein, Seth, and Emile A. Okal, 'Speed and Size of the Sumatra Earthquake', *Nature*, 434 (2005): 581–82
Stein, Seth, and Jerome Stein, *Playing Against Nature: Integrating Science and Economics to Mitigate Natural Hazards in an Uncertain World* (Chichester, 2014)
Strabo, *The Geography of Strabo*, vol. 7, Horace Leonard Jones trans. (London, 1930)
Stukeley, William, 'The Philosophy of Earthquakes', *Philosophical Transactions of the Royal Society*, 46 (1749–50): 731–50
Talwani, Pradeep, 'The Impact of the Early Studies Following the 1886 Charleston Earthquake on the Nascent Science of Seismology', *Seismological Research Letters*, 85 (2014): 1366–72
Titov, Vasily, Alexander B. Rabinovich, Harold O. Mofjeld, Richard E. Thomson and Frank I. González, 'The Global Reach of the 26 December 2004 Sumatra Tsunami', *Science*, 309 (2005): 2045–48
Tributsch, Helmut, *When the Snakes Awake: Animals and Earthquake Prediction* (Cambridge MA, 1982)
Tyabji, Azhar, *Bhuj: Art, Architecture, History* (Ahmedabad, 2006)
Vale, Lawrence J., and Thomas J. Campanella eds., *The Resilient City: How Modern Cities Recover from Disaster* (New York, 2005)
Waley, Paul, *Tokyo: City of Stories* (New York and Tokyo, 1991)
Walker, Bryce, and The Editors of Time-Life Books, *Earthquake* (Amsterdam, 1982)
Wallace, Robert E., ed., *The San Andreas Fault System: An Overview of the History, Geology, Geomorphology, Geophysics, and Seismology of the Most Well Known Plate-Tectonic Boundary in the World* (Denver CO, 1990)
Walpole, Horace, *The Letters of Horace Walpole*, vol. 2, Peter Cunningham ed. (Edinburgh, 1906)
Wang, Kelin, Qi-Fu Chen, Shi-hong Sun

and Andong Wang, 'Predicting the 1975 Haicheng Earthquake', *Bulletin of the Seismological Society of America*, 96 (2006): 757–95
Wegener, Alfred, *The Origin of Continents and Oceans*, 4th edn (New York, 1966)
Weisenfeld, Gennifer, *Imaging Disaster: Tokyo and the Visual Culture of Japan's Great Earthquake of 1923* (Berkeley CA, 2012)
Willmoth, Frances, 'Rumblings in the Air: Understanding Earthquakes in the 1690s', *Endeavour*, 31 (2007): 24–29
Winchester, Simon, *A Crack in the Edge of the World: The Great American Earthquake of 1906* (London, 2005)
Witze, Alexandra:
'Tsunami Alerts Fall Short', *Nature*, 516 (2014): 151–52
'Artificial Quakes Shake Oklahoma', *Nature*, 520 (2015): 418–19
'The Quake Hunters', *Nature*, 523 (2015): 142–44
Wright, Alan, ed., *Film on the Faultline* (Bristol, 2015)
Yeats, Robert, *Active Faults of the World* (Cambridge, 2012)
Earthquake Time Bombs (Cambridge, 2015)
Zeilinga de Boer, Jelle, and Donald Theodore Sanders, *Earthquakes in Human History: The Far-Reaching Effects of Seismic Disruptions* (Princeton NJ, 2005)

ACKNOWLEDGMENTS

Jamie Camplin of Thames & Hudson got me started on writing about earthquakes by commissioning a book on great natural forces, published as *Earthshock* in 1993. *Earth-Shattering Events*, too, owes its existence to him, before his retirement in 2015. I shall miss him as my supportive and enlightened editor for almost thirty years. Sincere thanks, Jamie. Sophy Thompson promptly took up the baton, and improved the manuscript. Kit Shepherd was a thorough and thought-provoking copy-editor. Poppy David worked enthusiastically on the illustrations. Celia Falconer, Aaron Hayden, Sarah Vernon-Hunt, Amanda Vinnicombe and especially Jen Moore gave exemplary service in the design and production of the book.

Among the hundreds of experts whose published work I have drawn on, I would like to thank Paul Cartledge for his help with the history of the Sparta earthquake, and Eric Force, Susan Hough and Seth Stein for their advice on the history of seismology and its current understanding (and lack of understanding) of earthquakes.

SOURCES OF ILLUSTRATIONS

Frontispiece US Geological Survey; **6** Giovanni Caselli; **24** Photo April 2016 Museum of Fine Arts, Boston; **42** The Print Collector/Alamy; **56** AP Photo/Armando Franca; **74** Courtesy Heritage History:http://www.heritage-history.com; **106** US Geological Survey; **124** Earthquake Research Institute, The University of Tokyo; **142** Tangshan Earthquake Museum; **160** Courtesy Professor Sudhir Jain; **174** Mark Pearson/Alamy; **190** DigitalGlobe via Getty Images; **208** US Geological Survey; **236–37** Martin Lubikowski, ML Design, London

INDEX

Abe, Shinzo 205
Aceh (Indonesia) 21, 177, 179–84, 187–8, 211
Aceh, sultans of 182
Adams, John 85
after the quake (Murakami) 226, 234
Aftershock (movie) 209–10
Ahmedabad (India) 161, 164, 168, 170, 172
Akutagawa, Ryunosuke 135
Alaska 13, 175, 188, 208, 213, 228
Aleppo 30
Aleutian Islands 129, 188
Algeria 12, 63
Algiers 63
Allah Bund (India) 163–4
Altez, Rogelio 76, 77
Ambraseys, Nicholas 224
Anaxagoras 27
Anaximander 31
Anaximenes 27
Andes (mountains) 15
Andong Wang 144
Anemospilia (Crete) 25
Aneyoshi (Japan) 196
animals (and earthquakes) 25–6, 45–6, 62, 143, 215, 217–18
Anjar (India) 163, 164–5
Ankara 15
Antakya—*see* Antioch
Antigua (Guatemala) 16
Antioch (Turkey) 16, 35–6, 37
Arana, Marie 78, 86
Argentina 75, 87
Aristotle 27, 32, 51
Armageddon (Israel) 17, 19, 30
Ashkelon (Israel) 30
Assam 104
Athens 15, 16, 32
Australia 66, 130, 176, 189
Ayodhya (India) 165, 166
Aztecs 25

Bam (Iran) 223, 229
Banda Aceh (Indonesia) 174, 179, 180, 181, 187
Bangladesh 176, 177
Barnes, Gina 227
Basilicata (Italy) 90, 92
Beach Boys, The 232
Beijing 15, 20, 143, 144, 145, 146, 148–50, 153, 155, 157, 210
Ben-Menahem, Ari 29
Bentley, Richard 50
Bharatiya Janata Party 165–6, 167–8, 171
Bhatia, Gautam 173
Bhuj (India) 21, 161, 163, 164, 165, 166–7, 170–71
Bible (Old and New Testament) 15–16, 18, 27–30, 31, 48, 51, 78
Bilham, Roger 10, 66, 84, 95, 109, 172, 224
Bina, Andrea 97
Blegen, Carl 18
Bolívar, Simón 20, 74–82, 86–7
Bolivia 20, 75, 87
Boston 13, 83, 89
Boves, José Tomás 86
Boxer, C. R. 61
Brady, Brian 214, 222
Branner, John Casper 116
Brazil 19, 57, 60, 61, 65, 72, 87
Britain/United Kingdom 8, 12, 41, 42–55, 60, 66, 68, 82, 84, 86, 87, 90, 98, 99, 103, 164, 166, 195, 218
British Geological Survey 43
Browning, Iben 221–2
Burma (Myanmar) 176
Burnes, James 163
Burton, William 102
Bushnell, David 86
Byerly, Perry 97

Cairo (Egypt) 15, 223, 224
Calabria (Italy) 89–90, 92
Calais, Eric 212
Calcutta 172
California 7, 13, 15, 95, 96, 104, 106–23, 131, 175, 211, 212, 213, 216, 217, 220, 221, 222, 230, 231–3
California Institute of Technology 123, 228
Callao (Peru) 63
Calvert, Frank 30

Campanella, Thomas 211
Canada 8, 83, 176, 204, 210
Candide (Voltaire) 59, 70
Cao Xianqing 143, 144
Caracas 15, 20, 74–8, 79, 80–81, 83, 84, 85, 86, 87, 89, 211
Carchemish (Syria) 30
Caribbean 12, 17, 44, 66, 76, 83
Cartagena (Colombia) 78, 79, 80, 82, 86
'Cartagena Manifesto' (Bolívar) 80–81, 86
Cartledge, Paul 32
Caruso, Enrico 111, 112
Cascadia subduction zone 204
Catania (Italy) 90
catfish prints—see *namazu-e*
Cecchi, P. F. 97
Chang Qing 154, 155
Charleston (USA) 13, 82, 83, 103–04
Chen, Beatrice 147, 152, 154, 158
Chen Zhu-Hao 154–5
Chile 11, 12, 13, 15, 75, 87, 175, 188, 196, 199, 233
China 8, 12, 13, 16, 20, 21, 25, 37–9, 129, 139, 141, 142–59, 161, 167, 181, 204, 209–10, 218
Christchurch (New Zealand) 22
Cicero 31
Clancey, Gregory 102, 128, 131
Cline, Eric 18–19
Cluff, Lloyd 123
Cocos Islands 176
Coen, Deborah 57, 119
Colchester (UK) 12
Colombia 12, 20, 75, 78, 87
Colombo 21, 178, 179, 186
Colosseum (London) 57, 58
Colosseum (Rome) 6, 15, 33, 36, 227
Communist Party (of China) 20, 39, 143, 153, 154, 158, 210
Concepción (Chile) 11, 13
Condon, Emmet 114, 115, 117, 118, 120
Constantinople—*see* Istanbul
Cook, James 68
Cultural Revolution (China) 21, 141, 151–2, 156, 157, 159

Dalai Lama 197–8
dams 16, 30, 163–4
Darwin, Charles 11, 13, 22, 77
Davison, Charles 11, 46, 50, 67
Delhi 15, 161, 166, 170, 172, 173
Delpeche, Louis 77
Deng Xiaoping 21, 145, 146, 149, 155, 156, 157, 158
Denver 7, 8
Deraniyagala, Sonali 178
Dewey, James 49
Dholavira (India) 25, 162
di Tiro, Hasan 183
Diamond, Jared 17
Diaz, José Domingo 75–6, 77, 78
Dickens, Charles 58, 73
Dikötter, Frank 159
Diodorus Siculus 31–2
Dominican Republic 212
dos Santos, António Ribeiro 68
Dover 11, 50
Drews, Robert 31
Durant, Will 17
Dutthagamani 185
Dwyer, Jeremiah 110, 111

earthquake:
  economics 7–10, 13–14, 65, 113, 115–20, 121, 137, 138–9, 150, 157–8, 166–9, 181, 189, 197, 201, 211, 212, 213, 221, 224, 230
  epicentres 23, 39, 43, 55, 65–6, 84, 96, 98, 103, 131, 147, 161, 170, 172, 175, 176, 179, 181, 187, 189, 191, 213, 216, 223, 225, 226
  fatalities 12, 16, 17, 20, 21, 22, 36, 38, 39, 62, 63, 65, 69, 76–8, 93, 102, 112, 114–15, 121, 125, 130, 144, 147, 151, 161, 162, 164, 167, 172, 175, 177, 178, 179, 180–81, 188, 197, 215, 223, 225, 229
  insurance 115, 118, 122–3, 221
  intensity scales 90, 95–6, 229–30
  intraplate 83, 162, 223, 230
  magnitude scales 13, 95–6
  prediction 38, 47, 79, 84, 128–30, 131, 143–6, 212–23
  proofing (of buildings) 22, 33, 72, 100, 101, 120–21, 123, 158, 172, 173, 204, 224, 225, 227–30, 231, 233
  storms 19
  waves 54–55, 62, 89, 91, 94, 98, 104, 107, 131, 195, 198, 216

Earthquake (movie) 232
earthquakes:
- Alaska (1964) 175
- Anjar (1956) 164–5
- Ansei (1855) 39–40, 98, 102, 128, 130
- Antioch (526) 36
- Bam (2003) 223, 229
- Cairo (1992) 15, 223–4
- Calabria (1783) 89–90, 92
- Caracas (1812) 20, 74–87, 89, 211
- Chile (1835) 11
- Chile (1960) 13, 175, 188, 196, 199
- Chile (2010) 13, 196
- Christchurch (2010) 22
- Fort Tejon (1857) 29
- Great English (1884) 12
- Great Kanto (1923) 16, 20, 22, 41, 60, 62, 102, 124–41, 151, 173, 191, 205–07, 211, 213, 216, 218, 224, 226–7, 231
- Great Neapolitan (1857) 33, 88, 90–95, 100, 211
- Gujarat (2001) 21, 160–72, 234
- Haicheng (1975) 143–4, 145, 156, 217, 221
- Haiti (2010) 22, 211–12, 224
- Haiyuan (1920) 129, 153
- Hayward (1868) 220
- Jericho (1927) 16
- Kashmir (2005) 162
- Kobe (1995) 206, 225–6, 233, 234
- L'Aquila (2009) 214–15
- Lisbon (1755) 14, 16, 19–20, 50, 54, 55, 56–73, 77, 83, 89, 90, 131, 179, 195, 211, 218
- Loma Prieta (1989) 120–21, 122, 221, 227, 232
- London (1750) 12, 42–55, 62, 67, 89
- Long Beach (1933) 121–2
- Meiji Sanriku (1896) 193, 195, 202
- Messina (1908) 105
- Mino-Owari (1891) 102, 105, 128
- Nepal (2015) 171, 172, 173
- Nepal-Bihar (1934) 14, 172
- New Madrid (1811–12) 82–4, 162, 221–2, 223, 229–31
- Niigata (1964) 216
- Northridge (1994) 212, 233
- Pompeii (62 or 63) 16, 34
- Port Royal (1692) 17, 44
- Rann of Kutch (1819) 163–4
- San Fernando (1971) 216, 228
- San Francisco (1906) 14, 22, 95, 104–05, 106–123, 125, 131, 138, 162, 175, 211, 231, 233
- Shaanxi (1556) 38, 153
- Sichuan (2008) 210
- Sparta (*c.* 464 BC) 31–2, 33, 218
- Sumatra-Andaman (2004) 14, 21, 174–89, 193
- Tangshan (1976) 16, 20–21, 39, 141, 142–59, 161, 173, 209–10, 211, 216, 217
- Tohoku (2011) 22, 189, 190–207, 211, 225, 226
- Tokyo-Yokohama (1880) 99, 100–01

East Pacific Rise 176
Ecuador 20, 75, 87
Edo—*see* Tokyo
Egypt 25, 31, 36, 223–4
Elala 184
elastic rebound theory 108–09, 215, 222
Enkomi (Cyprus) 30
Escape from L.A. (movie) 232
Essay on Man, An (Pope) 70
Euripides 18
Evans, Arthur 18
Ewing, James 99, 103, 131

faults (geological) 7, 8, 25, 108–09, 128, 129, 194–5, 202, 212, 214, 222, 223, 224, 228
Feng Jicai 151–2
Feng Xiaogang 209
Ferrari, Graziano 92, 95
Finland 66, 83, 195
fires 15, 20, 25, 27, 28–9, 36, 39, 53, 54, 57, 59, 63, 64, 65, 67, 77, 100, 103, 113–19, 120, 121, 125, 127, 128, 131, 132–5, 137, 138, 144, 150, 173, 179, 226, 233
fissures/ruptures 29, 53, 67, 78, 94–95, 96, 106–08, 132, 213, 215
Fitzgerald, F. Scott 232
Flamsteed, John 52, 54
Folkes, Martin 45
fracking 8
Fradkin, Philip 232
Franklin, Benjamin 52
Free Aceh Movement (GAM) 21, 179, 183, 184, 187

Fukushima (Japan) 191, 192, 198, 199, 202
Fukushima Daiichi nuclear power plant (Japan) 22, 190, 198–204, 226

Galle (Sri Lanka) 179
Gandhi, Mahatma 14, 160, 164, 165, 167, 169
Gandhi, Rajiv 185, 186
Gandhi, Sonia 165
'Gang of Four' 145, 146, 156–7
Germany 8, 103, 139, 204
Ghatak, Maitreesh 168–9
Gibbon, Edward 35
Gibraltar 65, 66
Gilbert, Grove Karl 107, 116
Ginza district (Tokyo) 100
Global Positioning System (GPS) 162, 212, 217, 223, 230, 231
Goethe, Johann Wolfgang von 71
Gould, Peter 59
Gould, Randall 132
Grateful Dead, The 232
Gray, Thomas 99, 103, 131
Greece 12, 15, 17–18, 25, 26, 30, 33, 36, 82, 219
Guatemala 16, 26
Gujarat 21, 25, 160–73, 211, 216

Hadfield, Peter 225
Hagia Sophia (Istanbul) 227–8
Haicheng (China) 143–4, 145, 146, 156, 217, 221
Haiti 22, 75, 87, 211–12, 224
Haiyuan (China) 129, 153
Hales, Stephen 52
Hambantota (Sri Lanka) 179
Hamm, Harold 10
Hammer, Joshua 140
Hansen, Gladys 114–15, 117, 118
Harrison, Frances 187
Harvey, Robert 79–80
Hattusas (Turkey) 30, 35
Hawaii 188
Hayden, Edward Everett 104
Haydn, Joseph 57
Hearn, Lafcadio 193
Helots 31–2
Herculaneum 34, 57
Herod the Great 16, 27, 35
Himalaya (mountains) 162, 172, 175
Hirohito 125, 132, 138
Hiroshima 13, 57, 59, 63, 198
Hollywood 232–3
Homer 17, 26
Hong Kong 15
Honjo district (Tokyo) 133–4, 135–6, 137
Hooke, Robert 51–2, 54
Hoover, Herbert 147
Hough, Susan 10, 66, 83, 84, 95, 109, 144, 212, 217, 219, 227
Hu Keshi 146
Hua Guofeng 155, 156–7, 158
Humboldt, Alexander von 77, 78, 83, 85
Hyndman, Jennifer 184, 186

Iitate (Japan) 201
Imamura, Akitsune 124, 126, 128–9, 130–1, 134, 194–5, 213
Incas 10
India 12, 14, 21, 25, 36, 96, 104, 160–73, 175, 176, 177–8, 181, 184, 185, 189, 204
Indonesia 12, 174–7, 179–84, 185, 187–8
Indus civilization 25, 162
Inquisition 60, 63, 69, 71, 85
Iran/Persia 12, 16, 32, 223, 229
isoseismal map 94
Israel 12, 28
Istanbul (Turkey) 8, 15, 16, 19, 36, 227–8
Italy 12, 15, 25, 33–4, 52, 88–95, 97, 102, 105, 129, 214
Iyer, Pico 197

Jackson, James 14
Jaffna (Sri Lanka) 178, 185
Jakarta 15, 184
James, William 111–12
Japan 12, 13, 15, 20, 22, 25, 26, 39–41, 95, 97–103, 104, 105, 123, 124–40, 146, 181, 182, 189, 190–207, 210, 211, 213, 216, 224–7, 233, 234
Japan Meteorological Agency 26, 189, 191–2, 196, 199
Jefferson, Thomas 85
Jericho 15–16, 18, 27, 29–30
Jerusalem 19, 27, 37, 50
Jesuits 19, 61, 67, 68, 71–2
João V 61

José I 61, 64–5, 68, 71, 72
Julian the Apostate 35
Julius Caesar 35

Kan, Naoto 198, 204, 205, 206
Kanamori, Hiroo 123
Kant, Immanuel 14, 71, 218
Kanyakumari (India) 177–8
Karachi 15
Kashima (god) 24, 26, 40
Kashmir 162, 227
Kathmandu 171, 172
Kawabata, Yasunari 135–7
Kelin Wang 144
Kelvin, Lord 105
Kendrick, T. D. 50, 60
Kenya 176, 177
Kiyoshi, Muto 227
Klein, Naomi 14
Knight, Gowin 43
Knossos 17, 18, 19, 25, 30
Kobe 206, 225–6, 233, 234
Korea 20, 126–7, 151, 140, 206, 210
Krakatoa 175–6, 188
Kurosawa, Akira 125–7, 134, 135, 211, 226
Kutch (India) 21, 161–4, 166–7, 168, 170

Lachish (Israel) 30
L'Aquila (Italy) 215
Lawson, Andrew 108
LeConte, Joseph 103
Leibniz, Gottfried 69, 70
Li Yulin 149
Lienkaemper, Jim 220
Lima 15, 63, 214
Lisbon 15, 16, 19, 55, 56–73, 77, 89, 131, 179, 211
Livy 33
Lochbaum, David 199, 201, 203
London 11, 12, 42–55, 57–8, 60, 61, 62, 64, 65, 67, 68, 70, 91, 92, 89, 113, 134, 140, 167, 231
Long Beach (USA) 121–2
Los Angeles 13, 15, 29, 121, 123, 212, 213, 217, 232–3, 234
Lyell, Charles 92, 93
Lyman, Edwin 199, 201, 203
Lynch, John 75, 80

McConnell, Anita 92, 95
McWilliams, Carey 121
Madeira 66
Madison, James 85
Madrid 67
Mahdi, Saiful 180
Malagrida, Gabriel 71–2
Maldives 176
Mallet, Robert 33, 88, 90–95, 97, 100, 102, 104, 105, 108, 120
Managua (Nicaragua) 16
Manchuria 20, 139, 143
Manila 15
Mao Zedong 20–21, 38, 142, 145, 149, 151, 152, 154, 155–6, 157, 158, 210
Maori 26
Marina district (San Francisco) 121
Maxwell, Kenneth 68
Maya 10, 25, 26
Megiddo—*see* Armageddon
Mercalli, Giuseppe 96, 229
Merneptah 31
Messina (Italy) 105
Mexico 12, 25, 75
Mexico City 15
Michell, John 54–5, 89
Mid-Atlantic Ridge 176
Miletus (Turkey) 30, 35
Mill, John Stuart 14
Milne, John 14, 99–103, 105, 108, 128, 131, 214, 218
Minamisanriku (Japan) 197
Miranda, Francisco de 75, 78, 79–80, 85, 86
Missouri 13, 82–3, 221, 223, 230–31
Modi, Narendra 21, 166–71
Mongolia 25
Monteverde, Domingo de 78–9
Morocco 12, 63, 65
Murakami, Haruki 226, 234
Musson, Roger 197, 215, 219, 224
Mycenae 17, 18, 19, 30, 31

Nabataeans 36–7
Nallavadu (India) 178
namazu/*namazu-e* 24, 26, 39, 40–41, 128, 140, 218
Naples 15, 16, 33–4, 60, 89, 90, 91
Napoleon I 81–2, 86

Nehru, Jawaharlal 164, 165
Nepal, 12, 14, 161, 171, 172
Nero 16, 33, 34
New Madrid (USA) 82–4, 221, 223, 230–31
New Zealand 12, 22, 27
Newberry, John Strong 104
Newton, Isaac 43, 44, 48, 51, 54
Nield, Ted 177
Niigata (Japan) 216
Noda, Yoshihiko 205
Northridge (USA) 212, 233
Nur, Amos 18, 19, 20, 28, 30, 227

Odell, Kerry 120
Ofunato (Japan) 192, 193
Ogawa, Junya 207
oil industry 8–10, 161, 167, 168, 182–3
Okamura, Yukinobu 202
Oklahoma 7–10
Okuma (Japan) 200)
Omori, Fusakichi 98, 104–05, 128–31, 195
Osaka 15
Oshima, Noboru 132

Paice, Edward 59, 65
Pakistan 12, 161, 170
Palestine 15, 27–30, 31
Pallett Creek (USA) 220, 231
Palmdale (USA) 217, 218
Palmer, James 150, 151, 153, 154, 155, 156, 157, 158
Panama 87
Paris 60, 67, 70, 77, 81
Parkfield (USA) 213
Pasadena (USA) 123, 228
Patel, Keshubai 165–6
Perry, Matthew 40, 98
Peru 12, 20, 63, 75, 87, 176, 214
Petra (Jordan) 19, 36–7
Petrarch 33
Pickering, Roger 48
Pignataro, Domenico 90
Pilling, David 140, 196
Pisani, Elizabeth 180, 181, 184
plates (tectonic) 15, 66, 83, 95, 108, 162, 172, 175, 219, 223, 226, 230–31
Pliny the Elder 218
Pliny the Younger 34
Plume, Edward 110, 111, 112
Plutarch 218
Pombal, marquess of 19, 67–9, 71–2, 81
Pompeii 16, 34, 57, 58, 59, 60
Pope, Alexander 70
Port Royal (Jamaica) 17, 44
Port-au-Prince (Haiti) 22, 211–12, 224
Portugal 12, 19, 41, 56–73, 75, 81, 85, 211
Poseidon (god) 26, 31
Prabhakaran, Velupillai 185, 187
Premadasa, Ranasinghe 186
Price, Lance 171
Pylos (Greece) 30, 31

Qian Gang 145–6, 148–50, 151, 152–3, 154, 155, 209
Qi-Fu Chen 144

Rajapakse, Mahinda 187, 188
Ramesses III 31
Rashtriya Swayamsevak Sangh 164, 165, 166
Rebeur-Paschwitz, Ernst von 103
Reid, Harry Fielding 108, 109
Revi, Aromar 171, 172
Richter, Charles 13, 96, 121, 215
Rikuzentakata (Japan) 197
Rome 6, 15, 16, 33, 34, 35, 36, 85, 227
Romeo and Juliet (Shakespeare) 12
Rousseau, Jean-Jacques 70
Roy, Sanchari 168–9
Royal Society 12, 43, 45, 48, 49, 50–55, 67, 68, 89, 90, 91–2, 218
Rozario, Kevin 14, 72
Rumania 12
Russia 12, 204

Sagami Bay (Japan) 99, 128, 131, 213
Samuels, Richard 140, 203, 205, 206
San Andreas (movie) 232
San Andreas fault 7, 14, 15, 106, 108–09, 175, 194, 213, 216, 217, 220, 223, 231, 232–3
San Francisco 13, 14, 15, 22, 104–05, 106–21, 122, 123, 125, 129, 138, 179, 211, 213, 220, 221, 227, 231, 232, 233
Sanriku (Japan) 191, 193–4, 195–6, 202
Santiago 15
Satake, Kenji 191, 192, 193
Schmitz, Eugene 114

Schumpeter, Joseph 211
seiches 66, 195
Seidensticker, Edward 139, 140
Seikei, Sekiya 98
seismic bands 95
seismic electrical signal 219
seismic gap 128, 129, 171, 173, 220
seismographs/seismometers 7, 13, 19, 66, 89, 95, 96–8, 99, 101, 103, 105, 107, 129, 130–31, 175, 189, 191–2, 226
Seismological Society (of America) 120, 162
Seismological Society (of Japan) 99, 101
Sendai (Japan) 193, 197–8
Seneca 34
Shafter, Payne 107
shake table 228–9
Shakespeare, William 12
Shanghai 15
shear wall 228
Sherlock, Thomas 49–50, 55
Shide (UK) 99
Shi-hong Sun 144
Shinpei, Goto 137–8
Sieh, Kerry 220
Silicon Valley 14
Simpson, Edward 164, 165, 166, 170, 234
Sindh (Pakistan) 161, 162
Singapore 15, 167, 178
Smits, Gregory 40, 41, 194, 195, 203
Smolka, Anselm 212
Sodom (and Gomorrah) 28–9, 78
South Africa 176, 189
Southern Pacific railroad company 117, 118
Spain 20, 65, 66, 75, 78–9, 85, 86–7
Sparta 31–2, 33, 218
Sri Lanka 21, 175, 176, 177, 178–9, 181–2, 184–8, 189
Stanford University 107, 111, 112, 116, 122, 227
State Bureau of Seismology (China) 143, 145, 146, 156
Stein, Seth 109, 175, 201, 220, 222, 229–31
Strabo 28–9
Stukeley, William 43, 52–4, 55
subduction zones 15, 176, 189, 201, 204
Suetonius 33
Suharto 182
Sukarno 182
Sukarnoputri, Megawati 183
Switzerland 8
Syria 30, 35

Tacitus 34
Taipei 15
Taiwan 12
Tajikistan 216
Talwani, Pradeep 104
Tamil Nadu 177
Tamil Tigers (LTTE) 21, 179, 185–7
Tangier 63
Tangshan (China) 16, 21, 142–58, 161, 173, 209–10, 211, 216, 217
Taro (Japan) 196–7
Tecumseh 84
Teheran 15, 16, 223, 224
Thailand 176, 177, 181
Thales 27
Thiruvalluvar 177
Thucydides 27, 31, 32
Tianjin (China) 145, 151, 209
Tohoku 191, 192, 205, 206, 207
Tokyo 15, 16, 20, 22, 26, 32, 39–41, 59, 60, 96, 98–9, 100–01, 102, 103, 123, 124–41, 151, 173, 179, 191, 192, 200, 201, 204, 205, 206, 211, 216, 218, 224–6, 227, 231, 233, 234
Tokyo Electric Power Company 199, 200, 201, 202–5
Tokyo Imperial University 98–99, 101, 102–03, 124, 129, 130–31
Trajan 16, 36
Trans-Alaska Pipeline 208, 228
Tributsch, Helmut 217
Trincomalee (Sri Lanka) 185
Troy 17, 18, 19, 27, 30, 35, 63
tsunamis 14, 21, 22, 25, 27, 31, 39, 54, 55, 57, 63, 65, 67, 125, 174–89, 190–200, 201–02, 204, 205, 206, 207, 234
Turkey 12, 13, 16, 19, 35, 227
Turku (Finland) 66
Twain, Mark 232
Tyabji, Azhar 167

Ugarit (Syria) 30
United States of America 7–10, 12, 13, 14, 40, 81, 82–5, 89, 103–05, 106–23, 131, 137,

139, 147, 166, 180, 181, 198, 200, 204, 206, 212, 216, 221–3, 229–33
United States Federal Reserve 120
United States Geological Survey 7, 8, 9, 104, 213, 221, 222, 230

Vajpayee, Atal Bihari 166
Vale, Lawrence 211
VAN method (of earthquake prediction) 218–9
Venezuela 20, 74–82, 84–7, 211
Vesuvius, Mount 16, 34, 57, 59
Vishwa Hindu Parishad 164, 166
Vivekananda, Swami 177
volcanoes 16, 17, 27, 34, 51, 57, 59, 83, 99, 175, 188, 233
Voltaire 20, 59, 60, 69–71, 211
Voyage of the Beagle, The (Darwin) 6

Waley, Paul 134
Walpole, Horace 44, 45, 46–7, 49, 50, 52, 55
Warburton, William 48, 49, 55
Washington, DC 84, 200, 214
Weidenmier, Marc 120
Weisenfeld, Gennifer 138
Wells (UK) 11
Wesley, Charles 47
Wesley, John 48
Whiston, William 48, 50
Willmoth, Frances 52
Winthrop, John (IV) 89
Wolfall, Richard 63–4

yaodongs 38
Yokohama (Japan) 16, 20, 59, 99, 100–01, 125, 126, 131–2, 137, 138, 139, 206
Yoshida, Masao 199
Yudhoyono, Susilo Bambang 183, 184, 188

Zhang Heng 37–8, 97
Zhou Enlai 143, 145, 155